Blockchain Technology

Smart Innovation: Multi-Disciplinary Security and Communication Paradigm

Series Editor

Kanta Prasad Sharma

GLBAJAJ Group of Institutions, India

This series will encompass the topics of knowledge, intelligence, innovation, and sustainability. The aim of the series is to make available a platform for the publication of books on all aspects of single and multi-disciplinary research on these topics in order to make the latest results available in a readily-accessible form. Volumes on interdisciplinary research combining two or more of these areas is particularly sought. The series covers systems and paradigms that employ knowledge and intelligence in a broad sense. Its scope is systems having embedded knowledge and intelligence, which may be applied to the solution of world problems in industry, the environment, and the research community. It also focuses on the knowledge-transfer methodologies and innovation strategies employed to make this happen effectively. The combination of intelligent systems tools and a broad range of applications introduces a need for a synergy of disciplines from space science, war technology, business, and the humanities. The series will include edited collections, monographs, handbooks, reference books, and other relevant types of books in the areas of science and technology where smart systems and technologies can offer innovative solutions.

Blockchain Technology
Exploring Opportunities, Challenges and Applications
Edited by Sonali Vyas, Vinod Kumar Shukla, Shaurya Gupta and Ajay Prasad

For more information on this series, please visit: https://www.routledge.com/Smart-innovation-multi-disciplinary-security-and-communication-paradigm/book-series/CRCSIMDSCP

Blockchain Technology

Exploring Opportunities, Challenges, and Applications

Edited by
Sonali Vyas, Vinod Kumar Shukla,
Shaurya Gupta and Ajay Prasad

CRC Press is an imprint of the
Taylor & Francis Group, an informa business

First edition published 2022
by CRC Press
6000 Broken Sound Parkway NW, Suite 300, Boca Raton, FL 33487-2742

and by CRC Press
4 Park Square, Milton Park, Abingdon, Oxon, OX14 4RN

CRC Press is an imprint of Taylor & Francis Group, LLC

Library of Congress Cataloging-in-Publication Data

Names: Vyas, Sonali, editor.
Title: Blockchain technology : exploring opportunities, challenges, and applications / edited by Sonali Vyas, Vinod Kumar Shukla, Shaurya Gupta, and Ajay Prasad.
Other titles: Blockchain technology (CRC Press : 2022)
Description: First edition. | Boca Raton, FL : CRC Press, 2022. | Series: Smart innovation. Multi-disciplinary security and communication paradigm | Includes bibliographical references and index. | Summary: "This book is for anyone who wants to gain an understanding of Blockchain technology and its potential. The book is research-oriented and covers different verticals of Blockchain technology. It discusses the characteristics and features of Blockchain, includes techniques, challenges, and future trends, along with case studies for deeper understanding. Blockchain Technology: Exploring Opportunities, Challenges, and Applications covers the core concepts related to Blockchain technology starting from scratch. The algorithms, concepts, and application areas are discussed according to current market trends and industry needs. It presents different application areas of industry and academia and discusses the characteristics and features of this technology. It also explores the challenges and future trends and provides an understanding of new opportunities. This book is for anyone at the beginner to intermediate level that wants to learn about the core concepts related to Blockchain technology"-- Provided by publisher.
Identifiers: LCCN 2021050734 (print) | LCCN 2021050735 (ebook) | ISBN 9780367685607 (hbk) | ISBN 9780367685584 (pbk) | ISBN 9781003138082 (ebk)
Subjects: LCSH: Blockchains (Databases)
Classification: LCC QA76.9.B56 B562862 2022 (print) | LCC QA76.9.B56 (ebook) | DDC 005.74--dc23/eng/20211214
LC record available at https://lccn.loc.gov/2021050734
LC ebook record available at https://lccn.loc.gov/2021050735

ISBN: 978-0-367-68560-7 (hbk)
ISBN: 978-0-367-68558-4 (pbk)
ISBN: 978-1-003-13808-2 (ebk)

DOI: 10.1201/9781003138082

Typeset in Times
by Deanta Global Publishing Services, Chennai, India

This book is dedicated to family and friends …

Contents

Preface

We are at a beginning of a novel uprising in terms of technology. This uprising has started with economic revolution on the Internet and an alternative currency called Bitcoin, which is dispensed and sponsored by a vital authority instead of being automated in terms of achieving consensus amongst networked users due to its accurate uniqueness. Bitcoin and blockchain technology are a mode of decentralization, which is the next key disrupting technology and worldwide computing paradigm following the mainframe, personal computers and the Internet in addition to social networking/mobile phones. Blockchain is an insurrection encouraging a new world without any middlemen. Theoretically, it is an unchallengeable and tamper-proof distributed ledger of all transactions across a peer-to-peer network. With this book, you will get to grips with the blockchain network to build real-world projects. As you make your way through the chapters, you will be able to understand the major challenges that are associated with blockchain ecosystems, such as scalability, integration and distributed file management. By the end of this book, the chapters address common issues in the blockchain ecosystem. The primary authors in this book are from the Middle East and North Africa, United Arab Emirates, India, Malaysia, Qatar, Oman and Vietnam, and have covered both technical and local challenges related to blockchain technology. The book targets the opportunity, challenges and applications of blockchain technology and is a perfect platform for all experts in research and the academic community.

Preface

Acknowledgements

It is a privilege to introduce the book *Blockchain Technology: Exploring Opportunity, Challenges and Applications*, which focuses on emerging issues and challenges in blockchain technology. The book aims to provide a platform to inform researchers about emerging concepts related to the disciplines of blockchain technology. The book is a compilation of research-focused chapters, wherein the authors have discussed ideas ranging from contemporary technologies like cryptocurrency and concepts of mining to security and privacy issues in blockchain technologies, and so on.

We would like to extend our sincere gratitude to the authors of the chapters; without their dedication to research, this book would not have been possible. We would also like to thank the editorial committee, which has extended unfailing support for the conceptualization and finalization of this book. The launch of this book bears testimony to the zeal and commitment of the editors in providing a common forum for researchers to share their ideas and build upon them, adding to the process of knowledge creation. We are also very thankful to all our reviewers, who helped us to continuously improve the quality of the chapters and understand the different dimensions of the chapters. Very big thanks to the publishers, CRC Press, who have given us this wonderful opportunity, thanks to which all of us, editors/authors/reviewers, were able to come together for the conceptualization of this book.

We hope that academics, researchers and industry experts will find this book very useful as they set out to explore the fascinating world of advanced engineering, emerging technologies and an inspiring new form of digital currency.

Editor biographies

Sonali Vyas has served as an academician and researcher for around a decade. Currently, she is working as Assistant Professor (Selection Grade) at University of Petroleum and Energy Studies, Uttarakhand.

She has been awarded the National Distinguished Educator Award 2021, instituted by the International Institute of organized Research, which is registered with the Ministry of Micro, Small and Medium Enterprises, Government of India.

She was also awarded the Best Academician of the Year Award (Female) in the Global Education and Corporate Leadership Awards 2018.

Her research interests include blockchain, database virtualization, data mining and big data analytics. She has authored an ample number of research papers, articles and chapters in refereed journals/conference proceedings and books. She authored a book on *Smart Health Systems*, published by Springer. She is also an editor of *Pervasive Computing: A Networking Perspective and Future Directions*, Springer Nature and *Smart Farming Technologies for Sustainable Agricultural Development*, IGI Global. She acted as a guest editor in a special issue on Machine Learning and Software Systems in *Journal of Statistics & Management Systems* (Thomson Reuters). She is also a member of editorial boards and reviewer boards for many reputed national and international journals.

She has also been a member of the organizing committee, national advisory board and technical program committee at many international and national conferences, and has chaired sessions at various reputed international and national conferences.

She is a professional member of CSI Technology Group, the Institute of Electrical and Electronics Engineers, the Association for Computing Machinery-India, the Institute for Engineering Research and Publication, International Association of Engineers, the Internet Society, the Soft Computing Research Society and *International Journal of Engineering Research & Technology*.

Vinod Kumar Shukla is currently working with Amity University, Dubai, and United Arab Emirates in the capacity of Associate Professor – Information Technology and Head of Academics – Engineering Architecture Interior Design. He has more than 14 years of experience. He completed his PhD in the field of "Semantic Web and Ontology". He is an active member of the Institute of Electrical and Electronics Engineers. He has published many research papers in various reputed journals/conferences. He has also completed the General Management Programme from the Indian Institute of Management Ahmedabad (IIM-A). He is also a Cisco Certified Network Trainer. He has conducted many training programmes, including a training programme for the employees of Delhi Transco Limited, the Indian postal department in Megdoot Bhawan, Delhi, and Directorate General Resettlement, which is an inter service organization functioning directly under the Ministry of Defence. He has also written a case study on "Online Retail in UAE: Driven to Grow" and "Modi's Visit to UAE: Uncovering New Areas of Cooperation", which are hosted by European Case Clearing House, a case centre in the United Kingdom.

He received the Inquisitive Award in Unanimity-2020, organized and hosted by Amity University Dubai, United Arab Emirates. He has been awarded the Star Supporter Award 2019, which was presented by Manzil Center, Sharjah, United Arab Emirates; Her Highness Sheikha Chaica S. Al Qassimi and Dr. Ayesha Saeed Husaini, Director, Al Manzil Centre, Sharjah, United Arab Emirates awarded the Appreciation Certificate and Star Supporter Award 2019 to Amity University Dubai, received by Vinod Kumar Shukla for continuous support for community engagement and volunteerism. He has received Memento for Amity University Dubai from Mrs. Gloria Gangte, Deputy Chief of Mission, Embassy of India, Sultanate of Oman, for Guidance Seminar 2016.

Shaurya Gupta is currently working in the capacity of Assistant Professor (Selection Grade) with University of Petroleum and Energy Studies, Dehradun, Uttarakhand, India.

He has completed his PhD (Information Technology) from Amity University, Rajasthan. He has an MTech (Computer Science) from Jagannath University Jaipur and has over 11 years of experience in teaching, administration, liaison and coordination, student management, and research and analysis.

He has published various research papers and book chapters in reputed research journals and books and attended many national and international conferences as session chair and technical programme committee member. He is also an editor of many books and journal special issues.

His areas of interest are blockchain, delay tolerant network, wireless sensor network, ad hoc network, the Internet of Things and machine learning. Apart from research, he is involved in academics by means of preparing exercises, questionnaires and assignments for students at various levels, setting and marking assignments and tests, and assessing students' work for internally assessed components of qualifications at both undergraduate and postgraduate level.

Ajay Prasad, PhD Computer Science and Engineering (CSE), MTech CSE has more than 21 years of experience in faculty positions at reputed institutions with substantial industrial exposure. Currently, he is working as Senior Associate Professor at University of Petroleum and Energy Studies, Dehradun, India. He specializes in blockchain, computer architecture, simulation and modelling, network/data security principles, system programming, cloud systems, compilers and digital forensics. His research contribution is in fine-grained centralized monitoring in cloud computing and many allied areas of security and others. He has authored many research papers, a book on digital forensics, and many book chapters. He is a continuous contributor to society by means of cyber-awareness lectures and seminars to the public and government agencies like the Uttarakhand police department. He is a section editor for the prestigious journal *Space and Culture, India*. He has guided PhD scholars working in areas like Security in supervisory control and data acquisition systems, IEE802.11a-e, automated video surveillance, human aura and energy bio-field, the Internet of Things, etc. He is a life member of the Institution of Electronics and Telecommunication Engineers, the Indian Society for Technical Education and other prominent bodies in India. He has contributed to more than four different massive open online courses at the national level.

Contributors

V. Adarsh
Central University of Tamil Nadu
Thiruvarur, India

R Anitha
DST Cloud Research Lab
Department of Computer Science and Engineering
Sri Venkateswara College of Engineering
Erode, India

Martin Aruldoss
Central University of Tamil Nadu
Thiruvarur, India

Adarsh Kumar Arya
University of Petroleum and Energy Studies
Dehradun, India

Deepshikha Bhargava
University of Petroleum and Energy Studies
DIT University
Dehradun, India

Susheela Dahiya
University of Petroleum and Energy Studies
Dehradun, India

Rahul Das
UPES
Dehradun, India

Gayatri Doctor
CEPT University
Ahmedabad, India

Manik Gupta
Chitkara University
Punjab, India

Lam Oanh Ha
Thu Dau Mot University
Bình Dương, Vietnam

Sarfraz Hussain
Universiti Teknologi Malaysia
Johor, Malaysia

C. P. Igiri
Mountain Top University
Nigeria

Anubha Jain
IIS deemed to be University
Jaipur, India

M. Kar
Uludag University
Bursa, Turkey

Ridoan Karim
Monash University Malaysia
Subang Jaya, Malaysia

Keshav Kaushik
University of Petroleum and Energy Studies
Dehradun, India

Atiya Khan
G H Raisoni College of Engineering
Nagpur, India

Tinashe Mazorodze
BA ISAGO University
Gaborone, Botswana

Hazik Mohamed
Singapore University of Social Sciences
Singapore

Y Mohamed Sirajudeen
Vardhaman College of Engineering
Hyderabad, India

C Muralidharan
DST Cloud Research Lab
Department of Computer Science and Engineering
Sri Venkateswara College of Engineering
Erode, India

Kratika Narain
CEPT University
Ahmedabad, India

Van Chien Nguyen
Thu Dau Mot University
Bình Dương, Vietnam

Ajay Prasad
UPES
Dehradun, India

Neha Purohit
G H Raisoni College of Engineering
Nagpur, India

N Ramachandran
Indian Institute of Management Kozhikode
Calicut, India

Sakshi
Chitkara University
Punjab, India

Ashish K Sharma
G H Raisoni College of Engineering
Nagpur, India

Chetan Sharma
Chitkara University
Punjab, India

Durgesh M Sharma
G H Raisoni College of Engineering
Nagpur, India

Rewa Sharma
J.C. Bose University of Science and Technology
Faridabad, India

Sangita A. Sharma
Info Origin Technologies Pvt. Ltd
Gondia, India

Shamneesh Sharma
Poornima University
Jaipur, India

N K Shiju
Indian Institute of Management Kozhikode
Calicut, India

Vusumuzi Sibanda
BA ISAGO University
Gaborone, Botswana

Imtiaz Sifat
Monash University Malaysia
Institute for Management Research
Radboud University
Heyendaalseweg, AJ Nijmegen, Netherlands

A. R. Sowah
University of Ghana
Accra, Ghana

V. Srividya
VIT
Vellore, India

Salini Suresh
Dayananda Sagar College
Bangalore, India

Ruramayi Tadu
BA ISAGO University
Gaborone, Botswana

Miranda Lakshmi Travis
St. Joseph's College of Arts and Science
Cuddalore, India

B.K. Tripathy
VIT
Vellore, India

C. Udanor
University of Nigeria Nsukka
Nigeria

P Vinothiyalakshmi
DST Cloud Research Lab
Department of Computer Science and Engineering
Sri Venkateswara College of Engineering
Erode, India

1 Blockchain

Distinguishing Facts from Myths

N Ramachandran, N K Shiju, Salini Suresh and Anubha Jain

CONTENTS

DOI: 10.1201/9781003138082-1

1.1 INTRODUCTION

When an innovative technology grows by leaps and bounds, it also attracts an equal amount of speculation. The buzz word "blockchain technology" – a distributed, immutable ledger for recording transactions, tracking assets and building trust – is no exception to this phenomenon.

Blockchain is still in its exploratory stages. It is not just limited to cryptocurrency, contrary to popular perception, but extends beyond that. The fear factor associated with blockchain technology has also now spread into its use in cryptocurrency. Nevertheless, myths about distributed ledger technology (DLT) impede companies from leveraging its far-reaching potential and limit its positive impact on change. One cannot help but draw comparisons with the advent of cloud technology and the myths and misconceptions around this when it was touted to be just a bubble, susceptible and volatile. And yet, slowly, it gained momentum and captured the entire virtual world. Like the cloud process, blockchain has several positive aspects, but there is also a great deal of misunderstanding and scepticism about its purpose.

This study contributes to a greater understanding of this impending technology, which can no longer be kept at bay. And this chapter attempts to unravel the DNA of the most common misconceptions that persist about blockchain, distinguishing the myths from reality.

To achieve this objective, in Section 1.2, we give an overview of blockchain technology; in Section 1.3, we describe various myths about this technology and the facts about the myths; finally, in Section 1.4, we point to the conclusion of the study we have carried out.

1.2 OVERVIEW OF BLOCKCHAIN

In simple terms, blockchain is a distributed digital ledger; each transaction in the ledger is cryptographically signed and grouped as a block. When a new transaction happens, the new block is cryptographically connected to the previous block after proper validation, and it will be replicated to all nodes within the network. Once the new block is appended, we cannot modify the previous block, as shown in Figure 1.1 [1].

Since each block is cryptographically connected, it is tamper-evident, and it creates tamper resistance because we cannot modify the previous block once a new block is added. Moreover, the blockchain technology transaction happens without a central repository and a central authority or trusted third party such as a company, bank, government, etc.

1.2.1 Major Benefits of Blockchain

- As the blockchain uses only an appending ledger format, therefore easily tracks the entire transactions, and cannot be modified like traditional databases.
- Blocks in the blockchain are cryptographically secured; this ensures that the blockchain data cannot be tampered with.
- Since the ledger is shared with all nodes within the network, it ensures transparency, and it avoids a single point of failure.

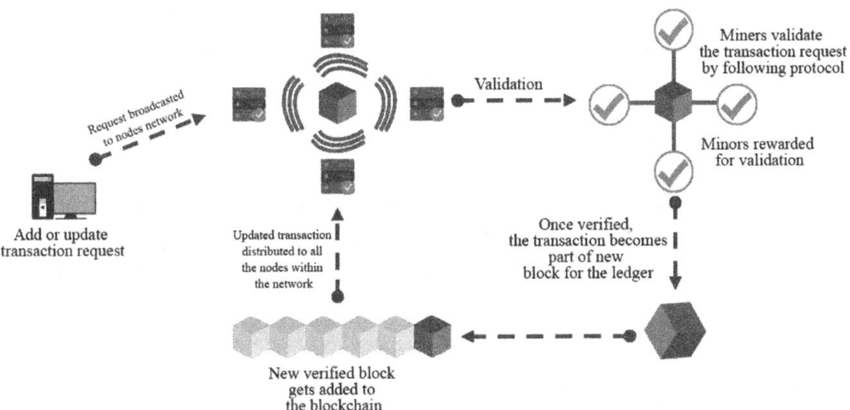

FIGURE 1.1 Overview of blockchain. (From Hileman, G. and Rauchs, M. *Global Blockchain Benchmarking Study*, Cambridge Centre for Alternative Finance, 2017. Available at: https://ssrn.com/abstract=3040224.)

- Blockchain technology works without intermediates; therefore, the transaction happens fast without charge or with a very nominal charge.

Some organizations tend to adopt new technology without understanding it, even if it is not necessary, due to overhype of the technology. Many organizations are afraid to implement the new technology due to misconceptions revolving around the technology, and blockchain has not been immune to this. The next section will reveal the facts of the myths revolving around blockchain technology.

1.3 MYTHS VERSUS REALITY

1.3.1 Blockchains Are Public

When we define blockchain as a global peer-to-peer network with transactions stored as blocks of chains across the network, we may think that it is a public platform and that since these blocks of nodes are spread out on a global scale, the data is accessible to all, and every activity happening in the network is visible to all. This is wrong; different security measures within the blocks and network, and even organizational-level accessible settings, are there.

There are different types of blockchain implementations, such as public, where data is visible to all in the network; private, in which data is structured and accessible to only designated people in the network; permissions, which are defined roles and accessibility levels with network usage regulations; and permission-less, where there are no regulations on network usage. [2]

1.3.2 Blockchain Is the Same as Cryptocurrencies

We have all heard the jargon of blockchain and cryptocurrencies like bitcoin. Most of us have a feeling that both are the same. But the fact is that cryptocurrency is an application of blockchain. Bitcoins use blockchain as the core technology for implementation. In the blockchain, everything is decentralized; we can also see it as peer-to-peer connections and transactions stored as blocks [3].

Bitcoins are cryptocurrencies for making electronic payments without involving banks or real cash by using virtual wallets. These currencies started to be used in 2009. Tracking of bitcoins is transparent, as it is a public ledger and uses blocks to store each transaction. These blocks are connected as a chain with timestamps of transactions. These ledgers are permanent, and no one can alter them. Not all cryptocurrencies use blockchain for their operations, but blockchain is ideal for cryptocurrencies [3].

1.3.3 Blockchains Can Never Be Tampered With

Most blockchain mythologies paint blockchain as an unhackable system, which is the main selling point of this technology. Unlike the traditional methodologies, the data is not stored in a single server but across the network as packets. Each transaction has also undergone different hash calculations, making it strong and resistant to tampering. But the fact is that no system is fully proof against hacking. If a group of nodes is malicious in a blockchain, the miners may rewrite the ledger, altering the smart contracts written or introducing new smart contracts.

The developers can take some steps to control tampering and make the blockchain private or permission, in which miner nodes will always be trusted nodes. Only these nodes should be allowed to perform actions like creating new nodes, setting up smart contracts, etc. There is no way trustless nodes can tamper with the state. The suggestion is to use the Proof of Authority (PoA) protocol [4]. This is a simple protocol to authorize signers to seal blocks or create a block by a voting technique. One more suggestion is to publish the hash status to an external node periodically. An auditor or an automaton can verify the previous hash and detect any tampering on nodes.

It is complicated to compromise the security features of blockchains, so the number of possible attacks aimed at this technology will be lower than with traditional systems. Still, we cannot say that blockchains can never be tampered with.

1.3.4 Free Blockchain

The most common myth about blockchain is its accessibility and cheapness of implementation. Blockchain requires high-end processing and computing power for the systems. It needs highly compact chips to hold data without compromising security. Algorithms and hashing techniques require the most powerful systems. These costs will fall on end-users or customers.

A disadvantage of decentralization and blockchain technology is that it is not free to use. The services and computing power must be paid for by the users. Users prefer centralized solutions because they will not be reminded of the cost of an operation on a regular basis, and the prices will be more hidden [5]. So, never believe that blockchain is free.

1.3.5 Smart Contracts Have the Same Legal Value as Regular Contracts

A smart contract is a set of code or some functions which will be triggered when specific conditions are met [6]. Today, smart contracts are prevalent and useful in every domain where the Internet of Things and artificial intelligence are utilized. When we consider an application to order online groceries when the stock is low in our refrigerator, some automated smart contracts will be triggered when the stock goes below the set level in this scenario. These may even involve financial transactions. There may be chances of reversals, too. If a customer files a case for the reversal of payment and takes this to court level, they may have an advantage in the eyes of the law, as smart contracts are not valid legal contracts. We can only present smart contracts as proof of transactions and say that they were executed because some conditions were met in the system. But, the customer may benefit from the bliss of ignorance due to the complexity of the system.

1.3.6 Blockchain Is Going to Change the World

Blockchain technology promises many innovative notions and advantages over traditional methods, resulting in people relying on this myth. The Bank of England's mounting interest in cryptocurrencies and other news on bitcoin acceptances adds fuel to this myth. A closer look at these shows that most exaggerated ideas are impractical and may turn out to be a bungling digital makeover of the conventional ledger system. Blockchain may have the potential to conquer the technological pathways, but the world has to wait for a flawless implementation to achieve this goal [7].

1.3.7 There Is Only One Blockchain

It is a common misunderstanding that blockchain is only one technology in distributed ledger architecture. But, there are other technologies, like Tangle, Hashgraph, etc., with the name of blockchain, which have a different type of implementation for the peer-to-peer distributed methods. When we consider the bitcoin blockchain network, the question has relevance, and there is only one bitcoin blockchain network. Studies are ongoing to scale up the blockchain network by including only financial-related or similar networks and making a more sustainable market [8].

1.3.8 Blockchain Is a Cloud-based Database

The most common blockchain myth is that everyone considers blockchain as a cloud-based database. But in reality, it is not a cloud-based application. Blockchain-based applications have to be installed on the client computer, and this should be connected to the internet. This computer must have a high speed and robust internet connectivity to ensure better chain strength. Any computer running this application is one of the nodes in blockchain and can store ledgers.

There may be physical files stored in the server in cloud-based applications, such as spreadsheets or documents. But, blockchain stores keep the records as proof of existence. This means that proof of existence clearly names a file or document, but it will not showcase the actual document [9].

1.3.9 Blockchain Is Used for Nefarious Purposes

When a new technology is introduced, users will always go through this kind of doubt. Many people do not have in-depth knowledge of how distributed, peer-to-peer, ledger-based technologies work. So, it is easy to be convinced and assume that they are used for wicked activities. But once we realize the facts in detail, our assumptions may change completely [10]. We can consider the scenario of cryptocurrencies themselves. They are more traceable than actual physical cash, as they rely on the digital network with immutable ledger-based architectures. Now, even government authorities are using blockchains in many of their crime tracking applications.

1.3.10 Blockchain Is Only Used for Cryptocurrencies

A common myth about blockchain is that it is only used for cryptocurrencies. But, this is not the case; blockchain is the building block of some cryptocurrencies, and the cryptocurrencies can use blockchain features for their implementation. All the mainstream domains can adopt the features of blockchain and make the applications more dominant.

For example, as well as the financial domain, the healthcare domain can utilize the advantages of blockchain, whereby patient data can float between institutions as nodes. The transactions on patient data can be kept as ledger activities (Figure 1.2).

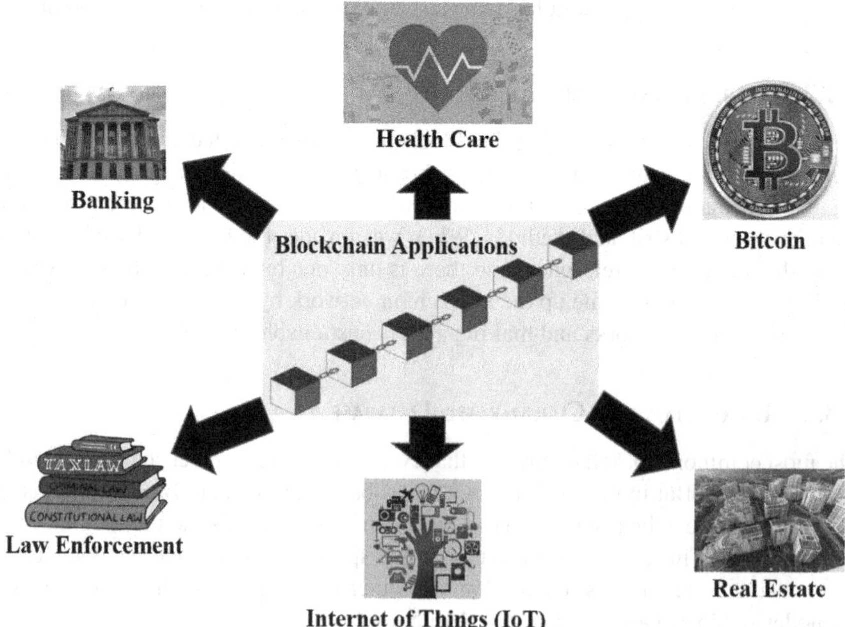

FIGURE 1.2 One of the many forms of blockchain technology is cryptocurrency.

1.3.11 Blockchain Is "Trustless"

In blockchain, trust level can be minimized but cannot be avoided. At least the base-level cryptography must be under trust. Permissioned blockchain can be the best minimum-level blockchain implementation if configured correctly. In this case, blockchain enables participants to validate transactions and verify them independently. So from this, we can say that blockchain can reduce or minimize the deed of trust. Blockchain can be used to integrate knowledge and instructions for a broad variety of applications as an immutable and tamper-proof public records archive for documents, contracts, properties and assets. Smart contracts, for example, are automated, self-executing activities in agreements between two or more parties; multi-signature transactions, on the other hand, necessitate the use of multiple signatures.

Under these scenarios, the benefit of blockchain is that it eliminates the need for a trustworthy third party (such as a notary) and enforces the execution of instructions via a cryptographic code, protecting participants from fraud and lowering management costs. There are substantial benefits of automation, transparency, auditability and profitability [11].

1.3.12 Blockchain Is a "Truth Machine"

Blockchain is also subject to Garbage In Garbage Out (GIGO) when it uses third-party non-native applications or external inputs. These inputs may go directly to the network as a regular entry, as blockchain cannot assess whether a third-party entry is valid or not. It can even trigger smart contracts, which may change the entire system. So, there should be a trusted third party who can validate the external inputs before inserting them into the existing blockchain.

1.3.13 Only Leading Developers Can Create Their Blockchains

The bitcoin blockchain is open source and so available for forking. Anyone who has access to the internet can fork this version and develop their own version. We see thousands of cryptocurrencies in the market, minor variations of bitcoin in some form. For an extended period, only leading developers were given access to the blockchain code, and thus, it was available only to them. But now, the open-source code is available, and developers can use it.

1.3.14 Blockchains Can Be Used for Anything

Blockchain is an instrumental technology, and one can think of many real-time convenient scenarios. It could be suited to, for instance, connecting students to blockchain and tracking their studies or connecting a citizen in blockchain and tracking their activities. It looks so apt in these situations, but it is not mature and is still in the very earliest stage. So currently, everything is in the experimental phase and not ready to take to this level. But at a later point in time, it may rule the world with loads of applications and chains.

1.3.15 Blockchain Transactions Are Anonymous

To an extent, blockchain transactions are anonymous, but not entirely. The transactions are saved with the public address of an individual's wallet. So, identity is safe. But now, many start-ups and organizations are working on blockchain and innovating many tools in this area. Suppose a person is trying to access the virtual public address of an individual's wallet with a real identity for some purpose. In that case, there is the possibility to list or track all the previous transactions linked with this identity.

1.3.16 Blockchain Can Power the Global Economy

Blockchain supporters hype bitcoins and other cryptocurrencies and believe that blockchain is the global economy's backbone. But this is not true. The size of the bitcoin network is not small; it is similar to that of the National Association of Securities Dealers Automated Quotations exchange (NASDAQ) network, but it is unlikely to take over any time soon.

1.3.17 The Blockchain Ledger Is Locked and Irrevocable

Today, financial transactions are linked to an institution, such as payment transactions and customer details being mapped to banks. Blockchain has the power of encryption and verifies transactions, which can even stop the system repaying or doubling payments. In this case, financial benefits are given in the form of coins only. However, the computational system is improving its functionalities, and chains or nodes may be revealed any day.

1.3.18 Blockchain Is Designed for Business Interactions Only

Blockchain started with business interactions only, but now the system is improving, and also, many institutions are working on this technology to get the most out of it. A user only needs internet connectivity to be part of a blockchain. So with this wider scope, many are trying to evolve suitable applications by using blockchain.

1.3.19 Blockchain Will Remove All Intermediaries

At present, the primary consumer of blockchain technology is bitcoin or similar cryptocurrencies. Bitcoin is primarily used in transactions; then came the myth that it can remove all intermediaries in traditional financial transactions and additional charges between the parties. In reality, it eliminates third-party mediators, but miners will act as mediators to validate and add the blockchain transaction. The only difference is that their name or functionalities may change in a more extensive scope.

1.3.20 Transactions on a Blockchain Are Automatically Immutable

It depends on the nature of the implementation of the blockchain. Blockchain enactment in bitcoin is immutable, but this involves a cost in computational capabilities. It involves untrusted third-party involvement in all parts of transactions, and hashing techniques should be sturdy enough to keep the data untampered with. It requires high-powered computational systems, and the implementation cost is high. Here, the auditors will notice any data change immediately, which may be cost-effective in implementation. As with any other system, blockchain is exposed to tampering.

1.3.21 Distributed Ledger Technology and Blockchain Are Synonymous

The term *DLT* also enters the discussion whenever we discuss blockchain. We consider blockchain also as a peer-to-peer distributed architecture. DLT is old technology, and blockchain uses its concepts in implementation. Blockchain is not DLT alone but has many more features [12].

DLTs and cryptocurrency were first suggested for the financial sector, including banks. Conversely, it was quickly discovered that DLTs are not restricted to trading in virtual currencies or goods but could be used to swap digital assets. The idea is that DLT encourages network users to pass and update information or documents, and that this is done in a trusted environment [13] (Figure 1.3).

1.3.22 Blockchain Is a Database

When a new technology arises, there is a common tendency to compare it with existing technologies. As blockchain is used to store data, we compare it with a database. A database is absolutely not a blockchain. Both are techniques used to store data. A database stores the data in a centralized server; the data will be shared with access

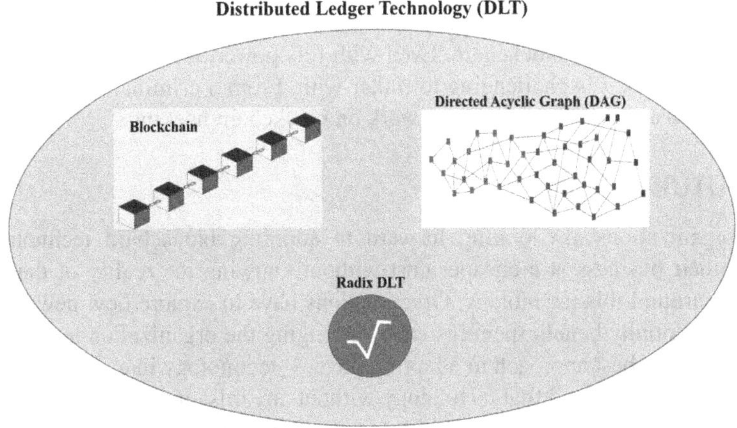

FIGURE 1.3 Blockchain is a subcategory of distributed ledger technology.

restrictions, and users with data manipulation rights can alter the database's data. In comparison, blockchain consists of distributed, immutable data connected over the network. The data may be visible or not visible to people in the chain depending on the implementation of blockchain.

1.3.23 Blockchain Coins Have Physical Equivalents

The term *blockchain coin* itself is not entirely true. By blockchain coin, the world refers to bitcoins. False news is being spread around, saying that bitcoins have physical equivalents, but this is wrong. The entire world relies on fiat currency approved by government regulations, which is a physical entity. But bitcoins are encrypted cryptocurrencies that only have a digital shape and are stored in individuals' crypto-wallets. Whenever a user performs a transaction, the number is updated in wallets and immediately reflected in the blockchain. As it is an immutable ledger, the transactions are permanent. So to participate in bitcoin transactions, a user should have a crypto-wallet and internet connectivity.

1.3.24 Blockchain Is a Fad

Blockchain is not a temporary fad. Countries like Dubai and China have already taken a giant leap in developing smart applications with blockchain. Dubai, for instance, may be one of the capitals of blockchain development. They are working on many initiatives to create the first city powered by blockchain. As a global technological leader, China is not lagging behind any other country working on the blockchain. They are even trying to create a new infrastructure for their economy.

1.3.25 Blockchain Augments Cyber Crimes

Tampering a blockchain is complicated, and a criminal may need the power of a million petaFLOPS for change in the chain, which is the power usually needed to create or delete a node in the blockchain. Even with this power, it may not be possible to alter the chain, as it is challenging to tinker with. From a criminal's perspective, it may be expensive and unprofitable to work on blockchain hacking.

1.4 FUTURE SCOPE

Many organizations are looking forward to adopting blockchain technology to develop their business at a cheaper cost without knowing the reality of the myths revolving around this technology. Organizations have to explore how new technology can potentially benefit them instead of changing the organization to fit the new technology. The best approach to adopting the new technology in an organization is to thoroughly understand that technology without any misapprehensions and implement it to reap its real benefit. We desire to carry out future research on the enabling and constraining factors of block chain technology.

1.5 CONCLUSION

Blockchain will have an effect on any field it crosses, from supply chains to human resources. Myths about what DLT is and how it operates, on the other hand, continue to dissuade companies from taking advantage of its much greater ability for progressive change. There is apprehension about blockchain technology. Nevertheless, myths about DLT impede companies from leveraging its far-reaching possibilities and limit its positive impact on change. The fear factor associated with blockchain technology has spread into its use in cryptocurrency technology. This chapter reveals the most common misconceptions that persist about blockchain technology. It elaborates the misconceptions about blockchain and its capabilities by taking a closer look at the reality. Furthermore, it reveals the shortcomings of blockchain, allowing a more accurate understanding of the technology.

REFERENCES

1. Yaga, D., Mell, P., Roby, N., & Scarfone, K. (2018). *Blockchain Technology Overview.* National Institute of Standards and Technology (NIST). Available at: https://doi.org/10.6028/NIST.IR.8202.
2. Mohan, C. (2019). State of public and private blockchains: Myths and reality. In *Proceedings of the 2019 International Conference on Management of Data SIGMOD 19,* Association for Computing Machinery, New York, NY, USA, pp. 404–411. DOI:https://doi.org/10.1145/3299869.3314116.
3. Garriga, M, Dalla Palma, S, Arias, M, De Renzis, A, Pareschi, R, & Andrew Tamburri, D. (2021). Blockchain and cryptocurrencies: A classification and comparison of architecture drivers. *Concurrency and Computation: Practice and Experience, 33,* e5992. https://doi.org/10.1002/cpe.5992.
4. Kaur, S., Chaturvedi, S., Sharma, A., & Kar, J. (2021). A research survey on applications of consensus protocols in blockchain. *Security and Communication Networks, 2021,* 6693731, 22 pages. https://doi.org/10.1155/2021/6693731.
5. Beck, R., Czepluch, J.S., Lollike, N., & Malone, S. (2016). *Blockchain: The Gateway to Trust-Free Cryptographic Transactions.* ECIS.
6. Bhargava, D., & Vyas, S. (Eds.). (2019). *Pervasive Computing: A Networking Perspective and Future Directions.* Springer.
7. Woebbeking, M.K. (2019). The impact of smart contracts on traditional concepts of contract law. *JIPITEC, 10,* 105. https://www.jipitec.eu/issues/jipitec-10-1-2019/4880.
8. Singh, S., Sharma, A., & Dr. Jain, P. (2018). A detailed study of blockchain: Changing the world. *International Journal of Applied Engineering Research, 13*(14), 11532–11539. ISSN 0973-4562 © Research India Publications.
9. Panwar, & Bhatnagar, V. (2020). Distributed ledger technology (DLT): The beginning of a technological revolution for blockchain. In 2nd International Conference on Data, Engineering and Applications (IDEA) (pp. 1–5), Bhopal, India https://doi.org/10.1109/IDEA49133.2020.9170699.
10. Sarmah, S. (2019). Application of blockchain in cloud computing. *International Journal of Innovative Technology and Exploring Engineering, 8,* 2278–3075. https://doi.org/10.35940/ijitee.L3585.1081219.
11. Gupta, S., Vyas, S., & Sharma, K. P. (2020, March). A survey on security for IoT via machine learning. In 2020 International Conference on Computer Science, Engineering and Applications (ICCSEA) (pp. 1–5). Gunupur, India: IEEE.

12. Sterley, A. (2019). Seven myths and misconceptions about blockchain debunked. *Accounting Blockchain Coalition*. Available at: https://accountingblockchain.net/seven-myths-and-misconceptions-about-blockchain-debunked/
13. Atzori, M. (2017). Blockchain technology and decentralized governance: Is the state still necessary? *Journal of Governance and Regulation*, *6*(1), 45–62. Available at: https://doi.org/10.22495/jgr_v6_i1_p5

2 Blockchain Technology and Cryptocurrency

Current Situation and Future Prospects

M. Kar

CONTENTS

DOI: 10.1201/9781003138082-2

2.1 INTRODUCTION

Human beings have developed by renewing themselves throughout history with the effort to meet their unlimited needs with limited resources. With the motivation created by the goal of maximum benefit and satisfaction, technology has developed rapidly and entered every field of life. At this time, blockchain technology and cryptocurrencies, which have entered the public agenda in recent years, have come to the fore as a powerful innovation and development and are being widely discussed in studies. Blockchain is a technology designed primarily for the safe and secure storage and management of valuable data. This technology is essentially a technical plan of a reliable data storage base that is stored in a decentralized and secure way and does not allow data to be deleted, lost or changed. Cryptocurrencies, on the other hand, are currencies and payment systems that use blockchain technology infrastructure without being connected to any centre. There are thousands of cryptocurrencies, the first to emerge and the best known being Bitcoin. Although efforts to combine the digital world and money go back to the 1980s, the subject gained general awareness with the emergence of Bitcoin at the end of 2008. With the widespread use brought by awareness and the increase in usage areas, it has come onto the agenda of the economic, social and political world and has been widely discussed.

The aim of this study is to examine cryptocurrencies that use blockchain technology, to reveal the current situation and to discuss its future. In the study, in order to enable readers and researchers who are not involved in the technical world to understand the nature of the subject, the technical definitions and analyses of technology are handled as superficially as possible, and a perspective more suitable for social scientists is used. Technical details are left to other chapters of this book and other studies in this field. The study is structured in ten different sections. After this short introduction, blockchain technology is discussed in the second section, and general information about its definition, features and structure is given. In the third section, the subject of crypto money is discussed, including definitions and history. The fourth section discusses whether or not cryptocurrencies are really money in view of the basic features and functions of money. In the fifth section, factors determining the value of cryptocurrencies are discussed. In the sixth section, the legal situation around the world is evaluated with some country examples. In the seventh section, the advantages and disadvantages of cryptocurrencies are discussed. In the eighth section, the current and potential uses of cryptocurrencies are discussed with various examples. In the ninth section, up-to-date data on the cryptocurrency market is shared. The tenth section is the conclusion. This section presents a general overview and evaluation considering the purpose of the study and makes some suggestions.

2.2 BLOCKCHAIN TECHNOLOGY

Blockchain technology is characterized as a promising technology within the framework of what it has revealed and promised, and it is predicted to play a critical role in the economic, social, political and military fields. This technology is a system designed to safely and securely store and manage valuable data such as money, identity and official documents. In blockchain technology, data is recorded not only by a centre or a group of centres but by everyone involved in the system. Blockchain enables users who do not trust each other to agree on unchangeable and irrefutable data without the need for a third party. In other words, what ensures trust in this system is not the relationships between individuals but the first set of rules of the system and the distribution of the record chain produced within these rules to everyone. Thanks to this technology, a transaction can be recorded without a central authority, so there can be independent secure communication between components. Blockchain technology makes it possible to record information or events irrevocably afterward together with the timestamp. When looking at the studies in the literature, there are three basic features of blockchain technology: Unchangeable, distributed and cryptographically secure. In addition, in the technical analysis of this technology, five different components are mentioned: Nodes, transactions, blocks, cryptography and consensus protocols. In a blockchain-based communication infrastructure, four universal security principles are provided: Authentication, non-denial, confidentiality and data integrity. In blockchain technology, there are three different network types: Private blockchain network, public blockchain network and consortium blockchain network (Gupta, 2017; Carlozo, 2017; Zheng et al., 2016; Crosby et al., 2016).

Blockchain technology, which came onto the global agenda with the introduction of Bitcoin in 2008, is an even older technology than cryptocurrencies and potentially offers advantageous solutions that can be applied in almost every sector. However, since the focus of this study is cryptocurrencies, one of the areas of use of blockchain technology, technical details on blockchain technology have been left to other chapters of this book and other studies in this field, and no further details will be given. In the following sections of the study, the subject of cryptocurrencies is brought to the fore.

2.3 CRYPTOCURRENCY

Having a dynamic structure throughout history, money has changed and transformed in line with economic, social and technological development. The adventure, which started with the use of materials such as seashells, beads and various stones, continues its journey with commodities money, representative money, fiat money and finally, digital money. Especially in the last quarter of the 20th century, as a result of this change, which has accelerated with the globalization process, cryptocurrencies show how far the concept of money has evolved today. Cryptocurrencies are discussed as the most up-to-date type of digital money. Money, as one of the leading subjects of the economic literature, has been subjected to different evaluations by different approaches and schools in terms of its definition, functions and features, and it has kept these discussions up to date with its development and transformation. Nowadays, at a time that concerns all actors in life, whether economic or not, cryptocurrencies are high on the agenda due to their advantages, promises and risks for the future. Whether cryptocurrencies are really money, whether they have the basic properties of money and whether they fulfil their functions, how their value is determined, what their legal status should be, and their current and potential usage areas are the issues that the literature is discussing at this point (Kumar & Smith, 2017; Griffith, 2017; Dierksmeier & Steel; 2018).

Cryptocurrencies were brought onto the agenda of humanity with Bitcoin, which was introduced in 2008 with a nine-page article titled "Bitcoin: Peer to Peer Electronic Cash System" published under the pseudonym Satoshi Nakamoto. In fact, the effort of the digital world to produce a digital form of money is not specific to Bitcoin. The article published by mathematician David Chaum in 1983 is one of the pioneering studies in this field (Chaum, 1983). David Chaum later founded the electronic money firm called "DigiCash" in 1990 based on this research. With the software offered by DigiCash, users were able to keep their money on their computers in a digital format called "eCash", signed by the bank cryptographically, and use this digital currency in any contracted institution in a confidential and secure manner without sharing information. Although this initiative went bankrupt in 1998 due to the insufficient number of users, the concepts and approaches it brought provided inspiration for future solutions.

Economic crises over time have shaken trust in the traditional monetary system, so interest in digital money is increasing day by day. The emergence of the digital currency Bitcoin in 2009, just after the 2008 crisis, is not considered a coincidence. According to research, the loss of the purchasing power of the US dollar over the years, the emerging new crises, and the search for new solutions have paved the way for the emergence of cryptocurrencies and their rapid growth in popularity. Bitcoin started mining and transfers by producing the first block (Genesis Block) on 3 January 2009. Bitcoin, which emerged in the environment of uncertainty in economic and financial systems and enables safe person-to-person payment without an intermediary institution, has rapidly grown in awareness and use with its innovative structure, simplicity and transparency. With the effect of the rapidly growing reputation of the pioneering cryptocurrency Bitcoin, many new cryptocurrencies have rapidly emerged with various changes to be used in alternative or more specific transactions. These cryptocurrencies, which emerged after Bitcoin, were called "altcoin", meaning alternative coin. Thousands of altcoins are currently in circulation in the market, and new altcoins are released every day (Nakamoto, 2008; Kumar & Smith, 2017; Griffith, 2017; Dierksmeier & Steel, 2018).

The use of cryptocurrencies is increasing day by day, and it has begun to attract the attention of all direct and indirect actors. The issue is being followed and investigated by all imaginable micro and macro actors, such as governments, regulatory agencies, central banks, financial institutions, academics and households. The first condition for a healthy discussion of important topics, such as making correct evaluations, taking necessary steps and taking necessary measures, is the correct and clear definition of cryptocurrencies. No clear, agreed definition has yet been established on this issue. In the study conducted by the Committee on Payments and Market Infrastructures, cryptocurrencies are defined as money that does not belong to any person or institution, is available electronically and can be transferred from person to person. The European Central Bank defines it as electronic money controlled by its own developers and used and accepted among members of a particular virtual community. The European Banking Authority, on the other hand, considers it as the digital representation of value that can be stored or sold, accepted as a medium of exchange by individuals and transferred electronically, but there is no central bank or public authority behind it, and it is not linked to a fiat currency. Cryptocurrencies, as the name suggests, are coins created with cryptographic encryption methods. Within the scope of the security provided by the encryption methods used, it is not possible to imitate, use and create illegal cryptocurrencies (Coron, 2006; Urquhart, 2016; Vigna & Casey, 2016; Garratt & Wallace, 2018).

2.4 COMPARISON OF CRYPTOCURRENCY AND MONEY

One of the important focuses of cryptocurrency discussions is whether cryptocurrencies are really money. In this respect, the easiest and most effective way is to look at cryptocurrencies in terms of the features and functions of money (Howells & Bain, 2008; Griffith, 2017; Vigna & Casey, 2016).

When money is mentioned, six different features come to mind: Portability, durability, divisibility, homogeneity, non-imitation and stability. In terms of portability, it can be said that cryptocurrencies can be transported more easily than money, since they are in a format that can be integrated into all mobile devices by using them in a digital environment. Money can deform over time due to its physical nature and use. However, cryptocurrencies are not physically minted and used. In this way, it can be said that cryptocurrencies are more durable than money. It can also be said that cryptocurrencies are more advantageous in terms of divisibility features. While cryptocurrencies have the ability to be divided into one cryptocurrency in a hundred million, this is not possible with money. From the point of view of the non-imitation feature, it is clear that imitation of cryptocurrencies produced with cryptographic encryption is more difficult than it is with money. In terms of the features of money, the most controversial issue is the ability to maintain its value over time, namely, stability. As stated before, there are many cryptocurrencies, and their values are very volatile. It is not yet possible to talk about a mature and stable process. The value of a cryptocurrency may clearly decrease or increase due to external effects. At this point, it is not yet possible to say that cryptocurrencies are superior to any money.

Money has three basic functions: A medium of exchange, a unit of account and a store of value. In daily economic life, money is widely accepted as a medium of exchange and used in return for the purchase of goods and services. Although it is a fact that cryptocurrencies are used more in the purchase of goods and services in the real economy than when they first appeared, this is far from saying that it is generally accepted. The fact that cryptocurrencies have not yet reached legal status and have high volatility limits the use of cryptocurrencies in the sale and purchase of goods and services by consumers and sellers. In this respect, it can be said that cryptocurrency has come a long way towards becoming a medium of exchange, but it is not yet a generally accepted medium of exchange. The second function of money is as a unit of account. This function enables the comparison of the relative values of goods and services by using money as a common unit of measure in the purchase and sale of goods and services. At present, the wide variety of cryptocurrencies and their very variable values make it very difficult to use them as a unit of account. In this respect, it can be said that cryptocurrencies cannot fulfil the function of being a unit of account. The third function to look at in the evaluation of cryptocurrencies is the store of the value function. People want to save. Money maintains its value over time, enabling people to use it for their desire to accumulate and keep wealth. For this, the value of money must be predictable. As mentioned before, the high volatility of cryptocurrencies prevents them from being used as a reliable tool for storing value. At this stage, those who anticipate that the value of a cryptocurrency will increase can buy it for investment purposes, but it is not possible to say that it is seen as a tool for value storage.

When the points discussed here in terms of both the features and the functions of money are evaluated, in general, it is quite difficult to say that cryptocurrencies are real money. Considering the distance it has travelled since the day it was first released, it is a fact that cryptocurrency has come a significant distance towards becoming money. It will continue to be debated whether cryptocurrencies fulfil the functions of money and whether they have these characteristics. However, what is indisputable is that cryptocurrency will move closer to the general definition of money as it provides stability and its widespread use increases.

2.5 VALUES OF CRYPTOCURRENCIES

One of the important discussions of the cryptocurrency agenda is how the value of cryptocurrencies is determined and what factors are affected. What should be stated primarily in this regard is that cryptocurrencies do not have value due to their nature; they gain value by being accepted by people. Just as the price of a normal good in the markets is determined depending on the supply and demand conditions, it can be said that the value of cryptocurrencies changes depending on the supply and demand conditions. However, it can also be said that prices can be affected by many different factors besides supply and demand, since there are no perfect competition conditions in the markets. Many factors, such as security, legal status, usage status, transaction speed, ease of use, costs, volume, supply, tax, crypto stock market, general perception, speculative and manipulative initiatives, news, etc., affect the value of cryptocurrencies. The most prominent among these factors are security, legal status and volume (Li & Wang, 2017; Garratt, & Wallace, 2018).

The most important factor affecting the value of cryptocurrencies can be expressed as security. The biggest reason for the cautious approach to cryptocurrencies is that it is a new technology, and society hesitates at the security point. A security vulnerability or the possibility of this will affect the value of money negatively, and every positive step taken at the security point will increase the value of the relevant cryptocurrencies. Another important factor affecting the value of cryptocurrencies is their legal status. While the adoption of cryptocurrencies by law will help rapidly increase their value, their evaluation as illegal will negatively affect their value. In addition, the taxation of these currencies is also linked to their legal status. Gaining legal status and subjecting them to excessive taxation may also negatively affect their value. The volume of cryptocurrencies in the market is also considered to affect their value. It is said that cryptocurrency markets are very volatile, which prevents people from investing. In an immature market, volatility is expected to decrease as the transaction volume increases over time. Another issue related to the volume of cryptocurrency is supply. As stated before, the value of cryptocurrency is related to the balance of supply and demand like a normal good. The supply of cryptocurrencies is usually determined from the beginning. This situation limits supply-related price movements. Since crypto markets are not yet fully known and have not reached large volumes, and the system has not been determined, their values are rapidly and greatly affected by news, perception operations, and speculative and manipulative initiatives.

2.6 LEGAL STATUS OF CRYPTOCURRENCIES

Official authorities are cautious about cryptocurrencies, which are met with great interest and considered relevant by innovative circles. The facts that cryptocurrencies are not dependent on any central authority, they can be easily used in international transfers, and price fluctuations are high make the authorities nervous. Especially in the first years, countries and organizations that have a say in the world economy have preferred to adopt tough attitudes. However, it can be said that these countries have loosened their approach over time, observing the increased interest and returns along with the risks. In general, it is said that developed countries are relatively more cautious and restrictive, and developing countries are more moderate with their efforts to turn the situation into an opportunity, but it is not possible to talk about a well-established and widely accepted approach to cryptocurrencies. Over time, three different approaches to cryptocurrencies have developed around the world: Banning them completely, allowing controlled development through auditing by giving restricted permissions, and staying neutral by not introducing any regulation or prohibition (Jackson, 2018; Lerer, 2019).

The United States is one of the countries whose steps at this point are being carefully followed due to their economic size and leadership in the volume of cryptocurrencies. Although cryptocurrencies are not recognized as legal in the United States, they have been removed from being perceived as a significant threat. Cryptocurrency exchanges are considered legal. Although cryptocurrencies are not legally accepted in Canada, cryptocurrency exchanges are legally evaluated with the registration requirement to the Financial Transactions and Reports Analysis Center of Canada. In China, cryptocurrencies are not accepted as legal payment instruments, and cryptocurrency exchanges are considered illegal. Cryptocurrencies are not accepted as a legal financial instrument in Russia. Although cryptocurrencies are not seen as a legal payment instrument in the United Kingdom, cryptocurrency exchanges are legal on condition of registration with the Financial Conduct Authority. The European Union recognizes cryptocurrencies legally, and regulations regarding cryptocurrency exchanges differ between member countries. In India, cryptocurrencies are not accepted as a legal payment instrument, but cryptocurrency exchanges are considered legal. South Korea is one of the countries that are brave about cryptocurrencies. The South Korean government supports the trading of cryptocurrencies and in 2018 passed a law allowing transactions with cryptocurrencies from real bank accounts. Japan, on the other hand, is shaping its policies in line with the goal of becoming Asia's cryptocurrency centre. In Japan, Bitcoin has been accepted as a legal payment instrument since April 2018. In addition, cryptocurrency exchanges can operate by registering with the Japanese Financial Services Authority.

2.7 ADVANTAGES AND DISADVANTAGES OF CRYPTOCURRENCIES

Whether cryptocurrencies will be accepted as money will be based on the evaluation of their returns and disadvantages as a result of rationality. Relevant market actors will evaluate the advantages of cryptocurrencies for themselves and compare them with their disadvantages. As a result, they will make a choice according to the total benefit of the use of cryptocurrencies. At this point, the prominent advantages and disadvantages of cryptocurrencies are very important (Hurlburt & Irena, 2014; Vigna & Casey, 2016; Troster et al. 2018; Smales, 2018; Weaver, 2018; Bruno & Gift, 2019).

2.7.1 Advantages

2.7.1.1 Security

A secure payment system is required in order to maintain healthy economic activities. Cryptocurrencies are based on blockchain technology that works with cryptology. Since mathematical methods and encryption are used in cryptology, information security is provided in this way, thus protecting both sender and receiver.

2.7.1.2 Limited Supply

In the traditional system, the source of a crisis is defined as short-sighted monetary policies or excessive expansionary fiscal policies. When cryptocurrencies start the first supply, it is clear how much will be supplied in total. The fact that the supply of cryptocurrencies is certain increases its predictability and eliminates the possibility of causing inflation.

2.7.1.3 Decentralization

Cryptocurrencies are not connected to a centre. Therefore, they are not subject to the rules of any state, bank or institution and cannot be directed. This advantage will enable the cryptocurrency market to maintain its stability once it becomes stable, regardless of external factors and political developments.

2.7.1.4 Low Costs

The fees and cuts made in money transfer reach very significant amounts for companies and become a major source of cost. It is possible to perform money transfer transactions with cryptocurrencies at much lower and sometimes zero costs.

2.7.1.5 Fast Transactions

Traditional payment systems can cause quite a waste of time when transferring money. Payment transactions or money transfers can sometimes take days. With cryptocurrencies, money can be transferred to the desired location in an unlimited and fast manner every day of the year.

2.7.1.6 Confidentiality

Even if transactions and amounts are open, people are hidden by their wallet addresses. Cryptocurrency does not have to be linked to any name or company. Thus, the tracking of money transfers is prevented, and the possibility of states seizing the money of individuals and institutions is eliminated.

2.7.2 Disadvantages

2.7.2.1 Extreme Technology

The cryptocurrency system is one of the most technological inventions of this age. For a large number of people who cannot easily give up their habits, using cryptocurrencies is very difficult due to the extreme technology involved. In addition, by their very nature, it will not be possible to compensate for possible errors due to irreversible transactions. This situation again discourages some people.

2.7.2.2 Digital Transformation

A major digital transformation is required for cryptocurrencies to be widely used in the system. It is obvious that after the system becomes operational, it will provide significant savings, but the costs of establishing the system are a deterrent. The most important deficit at this point is that there are not yet enough human resources in this area.

2.7.2.3 Private Keys

Access to cryptocurrencies is possible with the private keys given to the owner. It is the owner's responsibility to keep these private key passwords. Losing a private key means losing that cryptocurrency. It is not possible to reach that money by any other way or method.

2.7.2.4 High Volatility

Volatility can be defined as sudden price movements in the markets. The fact that the cryptocurrency market has a constantly fluctuating rate and a situation that is vulnerable to any kind of decline and increase at any time is shown as the biggest obstacle to the general acceptance of cryptocurrencies.

2.7.2.5 Lack of Legal Status

Cryptocurrencies are not accepted as legal payment instruments in many countries. The lack of legal status makes it impossible to claim a legal solution to the problems that may arise.

2.7.2.6 Illegal Affairs

Not knowing the identity of the parties in transactions and not being able to track the transfers provides a suitable environment for the use of cryptocurrencies in illegal activities.

2.8 CURRENT AND POTENTIAL USES OF CRYPTOCURRENCIES

With all the criticism and discussion, cryptocurrencies have rapidly expanded their usage areas over time, and although they have not fully shown their potential, they have presented important examples. It is expected that both the usage areas and the internal acceptance of cryptocurrencies will increase over time. In addition, the use of blockchain technology in areas such as online data storage, energy, financial sectors, insurance, supply chain, transportation, internet of things, foundations and donations, public practices, democratic practices, healthcare, real estate, digital identity, smart cities, intellectual property, law, education, inheritance, property rights and advertising will further support the use of cryptocurrencies. With all these developments, cryptocurrencies are expected to have all the functions and features of money and even to turn into much more than money over time. Some of the areas where cryptocurrencies are currently used and are expected to be used in the near future are discussed as examples in the following subsections (Bruno & Gift, 2019; Coron, 2006; Dierksmeier & Steel; 2018; Howells & Bain, 2008).

2.8.1 Tourism and Accommodation

The tourism and accommodation market, with its unique characteristics, is one of the markets that closely follow innovations in an effort to offer the best to consumers. For this reason, it is one of the sectors where cryptocurrencies have been most rapidly accepted and used. In the tourism sector, people travel to geographies where they do not even know the language and often need international money transfers to use in those countries. In this situation, travellers can make low-cost and fast money transfers with cryptocurrencies without the need for intermediaries. Cryptocurrency technology has created an easy and secure payment opportunity, especially for travel agencies that provide services over the internet and have an international character. A good example is the website Cheapair.com, which has accepted Bitcoin as a form of payment since 2013 when purchasing flight tickets, hotel reservations, car rentals and cruises.

2.8.2 Real Estate

The real estate sector stands out as one of the sectors that have rapidly begun to use cryptocurrencies. In general, everyone in society takes part in the real estate sector in one role or another as a buyer, seller or rental party. Real estate transactions are time-consuming, complex and costly transactions that many parties perform together and that require a reliable payment and recording system. If the process has an international character, the process becomes even more complex and difficult. With the use of blockchain and cryptocurrency technologies, it is possible to record the obligations of the parties in an unchangeable manner, to determine the accuracy of the information provided, to transfer the agreed payments safely and without cost, and to update the records in accordance with the new situation in an easy, fast and reliable manner. Propy.com and Mycoinrealty.com are the world's best-known international real estate sales sites that accept payments in cryptocurrencies to purchase real estate.

2.8.3 Education

The education sector has also pioneered the use of cryptocurrencies and is one of the markets that are expected to use cryptocurrencies more widely in the future. The role of educational institutions, especially universities, in the acceptance of cryptocurrencies is quite high in the eyes of society. Most of these institutions support this technological innovation as part of their mission. Educational institutions that prefer blockchain-based technologies in subjects such as diplomas, certificates, and transcripts accept cryptocurrencies in all kinds of payments. Accepting cryptocurrencies provides a significant convenience and cost advantage, especially for international students. Some universities in Switzerland, the United States, Germany and the Turkish Republic of Northern Cyprus have been pioneers in this field.

2.8.4 Charity and Donation Transactions

Among the areas where blockchain technology and cryptocurrency innovation provide important facilities are charities and donation campaigns. Corruption in these institutions and organizations leaves people hesitant to donate. With blockchain technology and cryptocurrencies, it can be easily observed where the money is coming from and where it is collected, how it is distributed and who has access. With the use of these technologies, donors can follow the process in a very transparent way, and if they wish, they can deliver the donation they want directly to the needy without the need for any intermediary. The World Food Programme currently provides good examples of these processes.

2.8.5 Supply Chain

It is generally accepted that the use of cryptocurrency in supply chain sectors such as transportation, logistics or material supply provides and will provide very important advantages. It is even possible to develop and use specialized cryptocurrencies for supply chains. The use of cryptocurrencies in the process accelerates the process, reduces costs and makes it easier to find solutions with indelible and unchangeable records. It is known that big companies such as JD.com, Alibaba, Provenance, Walmart and Toyota have tried the use of blockchain technology and cryptocurrencies in their procurement processes. There is also a cryptocurrency specially prepared for supply chains, called VET.

As stated before, there are many areas and markets where cryptocurrencies are used and are expected to be used. It is not possible to address all of them here. However, a few more examples can be given. For example, companies such as

Overstock.com, Crate and Barrel, Nordstrom and Whole Foods in the retail sector accept payments with cryptocurrencies. In the automotive industry, Tesla and Lamborghini sell some of their vehicles with cryptocurrencies. Solid Opinion, on the other hand, uses cryptocurrencies in digital publishing. Artworks are sold on the Bitpremier.com website by accepting cryptocurrencies. Uber makes cryptocurrency work with Facebook to be used in payments. On platforms such as ShareRing and Open Bazaar, buyers and sellers are securely brought together and paid with cryptocurrencies, enabling them to sell and buy the products they want (Bruno & Gift, 2019; Coron, 2006; Dierksmeier & Steel; 2018; Howells & Bain, 2008).

2.9 CURRENT INFORMATION ABOUT THE CRYPTOCURRENCY MARKET

According to the Coinmarketcap.com website, which is a reliable source where the cryptocurrency market can be followed, there are more than 4,000 cryptocurrencies as of February 2021. The total market cap of cryptocurrencies is over $1.6 trillion. When ranked according to total market cap, the top ten cryptocurrencies are Bitcoin (BTC), Ethereum (ETH), Binance Coin (BNB), Tether (USDT), Polkadot (DOT), Cardano (ADA), XRP (XRP), Litecoin (LTC), Chainlink (LINK) and Bitcoin Cash (BCH). When sorted by price, the top ten cryptocurrencies are renBTC (RENBTC), Bitcoin (BTC), Wrapped Bitcoin (WBTC), Bitcoin BEP2 (BTCB), yearn.finance (YFI), Maker (MKR), Ethereum (ETH), Bitcoin Cash (BCH), Compound (COMP) and Aave (AAVE) (Coinmarketcap.com, 2021, February).

Among these cryptocurrencies, the price of Bitcoin is around $50,500, the total market cap is over $1 trillion and the total supply in circulation is 18,637,106 BTC. The price of Ethereum is around $1,800, the total market cap is over $200 billion and the total supply in circulation is 114,775,384 ETH. The price of Binance Coin is around $250, the total market cap is over $39 billion and the total supply in circulation is 154,532,785 BNB. The price of Bitcoin Cash is around $660, the total market cap is over $$12 billion and the total supply in circulation is 18,663,200 BCH. The price of Tether is $1, the total market cap is over $34 billion and the total supply in circulation is 34,761,096,222 USDT. The price of Polkadot is around $36, the total market cap is over $33 billion and the total supply in circulation is 912,317,612 DOT. The price of Cardano is around $1, the total market cap is over $32 billion and the total supply in circulation is 31,112,484,646 ADA. The price of XRP is around $0.5, the total market cap is over $32 billion and the total supply in circulation is 45,404,028,640 XRP. The price of Litecoin is around $180, the total market cap is over $12 billion and the total supply in circulation is 66,544,865 LTC. The price of Chainlink is around $28, the total market cap is over $11 billion and the total supply in circulation is 36,635 YFI. The price of yearn.finance is around $36,000, the total market cap is over $1.3 billion and the total supply in circulation is 408,509,556 LINK. The price of Maker is around $2,370, the total market cap is over $2 billion and the total supply in circulation is 995,383 MKR (Coinmarketcap.com, 2021, February).

2.10 CONCLUSION

With the effort to keep up with rapid technological developments, the vehicles used by humanity are changing very quickly. Cryptocurrency has emerged as a product of this change and transformation. Like every innovation and invention, cryptocurrencies have come into life with their unique advantages and disadvantages. In this chapter, cryptocurrencies that use blockchain technology were examined, the current situation was described, and its future was discussed.

The results obtained in the study can be summarized as follows:

- It has been determined that cryptocurrencies cannot be defined as money, since they cannot totally fulfil the features and functions of money in their current situation. However, it has been concluded that the process they have gone through so far and the new gains of crypto coins in the future will approach the existing definitions.
- It was explained that the values of cryptocurrencies are shaped depending on the supply and demand conditions and are significantly affected by factors such as security, legal status, volume, and speculative and manipulative initiatives.
- It has been stated that cryptocurrencies do not yet have a widespread legal acceptance worldwide, but countries have loosened their initial rigid attitudes.
- It has been explained that among the many advantages and disadvantages of crypto money, the most prominent advantage is security, and the most striking disadvantage is volatility.
- It has been concluded that cryptocurrencies can be used wherever money is used, and further, they can offer new functions.

Not missing opportunities depends on being quick. In the light of risk assessments, it is necessary to implement regulations to protect crypto consumers, investors and the market. It is necessary to understand crypto money well in order to benefit from its gains and avoid its risks. Blockchain technology and cryptocurrency issues are hot topics for academic studies. Since cryptocurrencies have been in the market for more than 10 years, pioneering empirical studies should be done and compared with theoretical studies. In addition, new studies can be conducted to examine how the Covid-19 pandemic will have an impact on the acceptance of cryptocurrencies.

REFERENCES

Bruno, T.D., & Gift, L. (2019). How businesses can deal with cryptocurrency risks. *Intellectual Property & Technology Law Journal, 31*(3), 20–22.

Carlozo, L. (2017). Understanding blockchain. *Journal of Accountancy, 224*(2), 1.

Chaum, D. (1983). Blind signatures for untraceable payments. *Advances in Cryptology, 82*(3), 199–203.

Coinmarketcap.com (Accessed: 01.01.2021-02.20.2021).

Coron, J.S. (2006). What is cryptography?. *IEEE Security & Privacy*, *4*(1), 70–73.
Crosby, N., Pattanayak, P., Verma, S., & Kalyanaraman, V. (2016). Blockchain technology: Beyond bitcoin. *Applied Innovation*, *2*(71), 6–10.
Dierksmeier, C., & Steel, P. (2018). Crytocurrencies and business ethics. *Journal of Business Ethics*, *152*(1), 1–14.
Garratt, R., & Wallace, N. (2018). Bitcoin 1, bitcoin 2,....: An experiment in privately issued outside monies. *Economic Inquiry*, *56*(3), 1887–1897.
Griffith, K. (2017). *A quick history of cryptocurrencies BBTC-before bitcoin*. Dinero.
Gupta, S.S. (2017). *Blockchain*. Wiley.
Hirawat, A., & Bhargava, D. (2015). Enhanced accident detection system using safety application for emergency in mobile environment: Safeme. In Proceedings of Fourth International Conference on Soft Computing for Problem Solving (pp. 177–183). Springer, New Delhi.
Howells, P., & Bain, K. (2008). *The economics of money, banking and finance: A European text*. Pearson Education.
Hurlburt, F.G., & Irena B. (2014). Bitcoin: Benefit or curse?. *IT Professional*, *16*(3), 10–15.
Jackson, O. (2018). US or Swiss approach for eu crypto regulation? *International Financial Law Review*, *22*. https://search.proquest.com/scholarly-journals/us-swiss-approach-eu-crypto-regulation/docview/2017889182/se-2?accountid=17219
Kumar, A., & Smith, C. (2017). *Crypto-currencies: An introduction to not-so-funny moneys*. Reserve Bank of New Zealand Analytical Notes Series, No. AN2017/07, 2017.
Lerer, M. (2019). The taxation of cryptocurrency. *CPA Journal*, *89*(1), 40–43.
Li, X., & Wang, C. (2017). The technology and economic determinants of cryptocurrency exchange rates: The case of bitcoin. *Decision Support Systems*, *95*, 49–60.
Nakamoto, S. (2008). Bitcoin: A peer-to-peer electronic cash system. https://bitcoin.org/bitcoin.pdf
Purohit, R., & Bhargava, D. (2017). An illustration to secured way of data mining using privacy preserving data mining. *Journal of Statistics and Management Systems*, *20*(4), 637–645.
Smales, A.L. (2018). Bitcoin as a safe haven: Is it even worth considering? *Finance Research Letters*, *30*, 385–393.
Troster, Vi., Tiwari, A.K., Shahbaz, M. & Macedo, D.N. (2018). Bitcoin returns and risk: A general GARCH and GAS analysis. *Finance Research Letters*, *30*, 187–193.
Urquhart, A. (2016). The inefficiency of bitcoin. *Economics Letters*, *148*, 80–82.
Vigna, P., & Casey, J.M. (2016). *The age of cryptocurrency: How bitcoin and the blockchain are challenging the global economic order*. Picador.
Weaver, N. (2018). Risks of cryptocurrencies: Considering the inherent risks of cryptocurrency ecosystems. *Communications of the ACM*, *61*(6), 20–24. https://doi.org/10.1145/3208095
Zheng, Z., Shaoan, X., Dai, H.N., & Wang, H. (2016). Blockchain challenges and opportunities: A survey. *International Journal of Web and Grid Services*, *14*(4), 352–375. https://doi.org/10.1504/IJWGS.2018.095647

3 Security and Privacy Issues in Blockchained IoT

Principles, Challenges and Counteracting Actions

Manik Gupta, Shamneesh Sharma, Sakshi and Chetan Sharma

CONTENTS

DOI: 10.1201/9781003138082-3

3.1 INTRODUCTION

Internet of Things (IoT) technology has enabled straightforward communication, efficient work environments, associated improved living and hastened innovation. At the same time, cyber security has become the prime need of today's world, and even new age technologies like IoT, smart technologies and information technology infrastructure management are not untouched by it (K. K. and Sharma, 2013). Any sort of vulnerability in the exclusions may be precisely the type of attack vector that leads to cyber-attack on IoT-based technologies. There are lots of vulnerabilities and attack vectors in this technology, which makes it insecure.

The speedy growth of Blockchain technology has affected our daily lives in every way. A study by the research and analysis firm Gartner (Gartner, 2017) guesstimated that "Blockchain technology will add $3.1 trillion in business value by the Year 2030". In an alternative investigation, the comprehensive worldwide IoT market is anticipated to grow $457 billion by 2020 (Sharma et al., 2017), from $157 billion in the year 2016. IoT is steadily cultivating itself in grade and potential to become one of the most ubiquitous technologies at present. According to one more study by Gartner, there will be upwards of 20 billion connected devices by the end of 2020 (Kishore & Sharma, 2016), from around 8 billion at the end of 2017.

3.1.1 Blockchain Technology

Blockchain is a tamper-proof digital ledger for any kind of data generation and propagation. Any new data generated in this chain is known as a block. Whenever a fresh block is created, it is common to every computer in the Blockchain peer-to-peer network. To create a new block, the data is confirmed by every node in the network. After the new data is verified by at least the first five network nodes, the data is converted into blocks. Every block encompasses new records and the hash of the previous block. Blocks are chained together using these hashes, and previous data is conserved forever. The Blockchain network reunites every 10 minutes, recording all the blocks created within these 10 minutes. There has been various research on the Blockchain hypothesis. The concept of Blockchain was initiated with the publication of a white paper authored by "Satoshi Nakamoto", who presented a novel digital currency system. This paper showed a technique with the help of which payments could be transferred directly without any intermediate party like banks (Ahram et al., 2017). Since then, researchers have investigated an assortment of facets of this technology and executed it to design a variety of products for the digital world both within and beyond financial transactions. In Singh & Singh (2016) and Treleaven et al. (2017), the authors talked about the prospects of Blockchain in the financial industry and discussed the most popular product of Blockchain, the bitcoin. There are three logical layers in the Blockchain technology: The application, Blockchain and data layers.

Application Layer: The application layer acts as an interface between the user and the Blockchain infrastructure. It provides overlays for different protocols, application programming interfaces (APIs) for programmers and platforms for end users. It supports decentralized applications.

Blockchain Layer: The Blockchain layer is responsible for the digital integrity of the data received from the data layer. It performs activities like rumour detection, information tracing and peer-to-peer information exchange to ensure integrity. Due to these checks, the data received to the user at the application layer is trusted, traceable, decentralized and with faster dememorization. Hence, we can say that it provides a mechanism for establishing identities and trust (Figure 3.1).

Data Layer: The data layer is basically the information source where data dissemination and classification take place on real-time data streams.

3.1.2 Internet of Things

The IoT (Sharma & Kishore, 2017) can be contemplated as a conservatory of conventional wireless sensor networks (WSN) that makes entity-to-entity communication possible by utilization of radio frequency identification (RFID) and near field communication (NFC). The IoT is the inevitable upcoming era of the Internet, which has led many researchers to concentrate on it. There are lots of architectures accessible in the field of IoT. There are various usages of IoT in real-life scenarios, too. WSN (S. S. & Kishore, 2015) has created a landmark in the field of information technology and put it on the motherboard of every technology through its applications. With the evolution of this technology, it has played its role in the IoT. Typically, the layered structure of IoT can be described as follows (Gupta et al., 2020).

Application and Middleware Layer: To build an integrated security scheme, this architecture integrates the middleware and application layers together. It collects data from the network layer, which conducts operations such as ubiquitous computation and decision-making and essentially maintains the data in the database. In addition, across various implementations such as intelligent living, intelligent farming, remote health monitoring, etc., this awareness of artefacts is processed worldwide. The danger of dissemination between networks is greater than in Internet IoT applications (Suo et al., 2012).

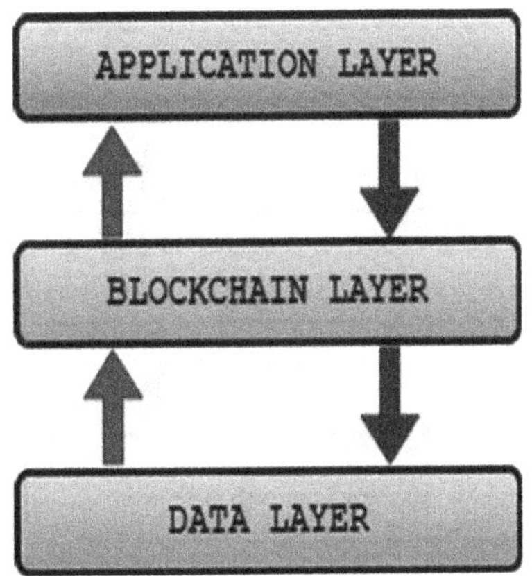

FIGURE 3.1 Three-layer architecture of Blockchain.

Network and Transmission Layer: This is generally referred to as the communication tier, which securely moves information from the physical equipment of the perception layer to the topmost tiers. Numerous heterogeneous network architectures that can be divided into local area networks (LAN), central networks or networks of access are enabled by this tier. Authentication and confidentiality of network data are the foremost security issues within this framework. This tier is extremely susceptible to attacks that can create network interference and have pervasive access to information on the perception layer; meanwhile, it holds a huge volume of data (Zhang & Wang, 2006) (Figure 3.2).

Perception Layer: This layer is the primary IoT design layer, which involves equipment such as global positioning systems, cellular sensor networks, radio frequency recognition, micro-electromechanical systems, sensors and/or nano-electromechanical systems. This involves object detection and tracking and the collection and acknowledgement of data from the physical devices or from the outside environment, which is thereafter translated and digitalized for data transmission to the communication layers for secure handling. The risks to safety on this layer are mostly from outside and physically affect or bypass the system program, such that nodes operate according to their desires and network knowledge is obtained; for the entity nodes involved, this layer offers unique protection services (Suo et al., 2012; Xiaohui, 2013). This has four main purposes: Authentication, security, sensitive data protection and risk analysis.

3.1.3 Blockchain-empowered IoT

In information technology, innovative technologies emerge frequently; very often, these technologies comprise the integration and amalgamation of different existing technologies to create a new technology. Blockchained IoT / Blockchain-enabled Internet of Things / Blockchain-based Internet of Things tries to maximize the effectiveness and efficiency of IoTs by implementing Blockchain algorithms at various

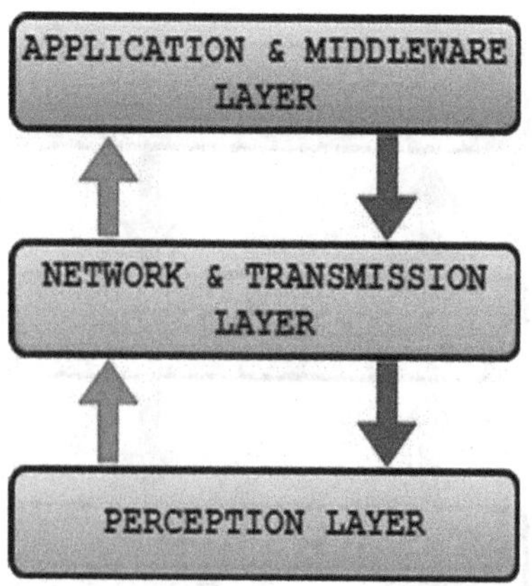

FIGURE 3.2 Architecture of IoT with respect to three layers.

layers. This enables the creation of secure networks, which provides autonomous, decentralized equivalents of traditional methods or tasks. It tries to achieve better security for the overall system. Some of the common mechanisms for securing IoT systems are registered "things", identity management, ownership validation and data history. Blockchained IoT can be implemented across all business domains (such as fintech, banking, hospitals and airlines) and technical verticals (such as Big Data, cloud and robotics). It can share data in a secure, private Blockchain to reduce the costs and complexities of existing business systems (Figure 3.3).

3.1.4 Layered Architecture of Blockchained IoT

The Blockchain technology manages a specific type of distributed ledger when implemented with IoT technology. The IoT technology is the successor to the WSN technology (Kishore & Sharma, 2016); the WSN was dependent on a LAN, whereas IoT has taken it to the broader concept of providing each node with Internet facility and making the process of data dissemination simpler (Sharma & Kishore, 2017) (Figure 3.4).

The integration of technologies needs five important aspects to be covered:

- Management of proof of ownership and identity in case of ubiquitous device management. This record of all the devices in the theme of identity, address and contract must be managed by the Blockchain ledger.

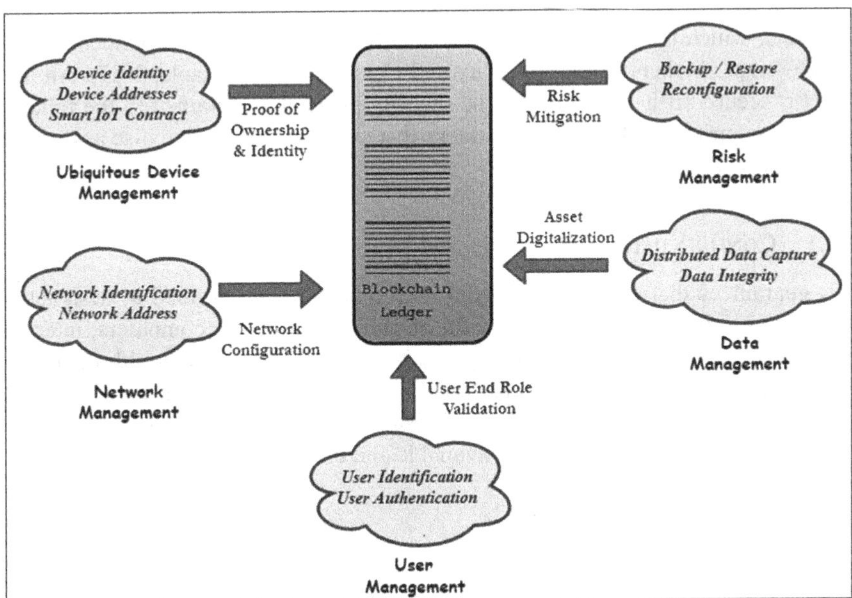

FIGURE 3.3 Proposed integrated model of IoT and Blockchain.

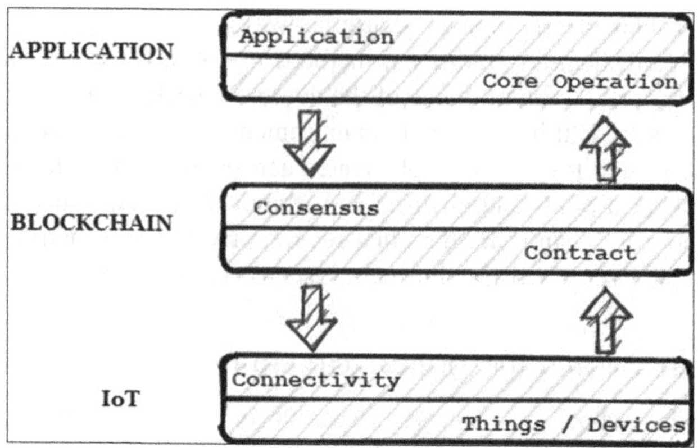

FIGURE 3.4 Layered view of Blockchain and IoT integrated architecture.

- Managing the network configuration is again a challenge, which has to be the responsibility of the ledger. In IoT, network identification is an important process and is the area most prone to attacks. When integrated with Blockchain, the process of network identification is properly recorded in the ledger, which makes it more secure.
- Most of the data is captured by the sensors in a network of IoT devices.

3.2 PRINCIPLES FOR SECURE COMMUNICATION IN IOT

Like general information security principles, conventionally, IoT also relies on the CIA triad, which is referred to as Confidentiality, Integrity and Authentication, or the AIC triad (Authentication, Integrity and Confidentiality) (Maple, 2017). An IoT object is secured if it fulfils any of the security principles explained in the following, and "security attacks" refers to attacks that successfully compromise any of the security principles.

3.2.1 Confidentiality (C)

This guarantees that exposure to confidential details can be limited to designated participants, which could be in the form of properties such as computers, internal artefacts, resources, and human beings. This idea is firmly in line with data protection and its classification policies, as it notes that access to infrastructure and resources can only be open in the case of a specific need, so that information limited to a single view can also be publicly available and usable to all network properties.

3.2.2 Integrity (I)

This guarantees the conservation of confidence, precision and coherence over the whole life cycle, i.e., that data or knowledge cannot be abused when processed on computers or floated from one end of the network to the other. The confidentiality of the data is ensured by authentication measures such as hashing, encrypted data or checksum. One very common approach for preserving data integrity is one-way hash measurement for such data collection, which should move in parallel with the initial post prior to transmission; If the computed hash does not match the data at the recipient's end, end-to-end IoT protection is maintained as the data is converted over the network.

3.2.3 Availability (A)

This guarantees that all the tools and data are rendered accessible to all the IoT applications that require them. This includes maintaining the hardware used in IoT systems, patching applications and optimizing the network to do this. However, performance is resolved by natural disasters and crises for preserving redundant processes. The failure resistance, cluster availability and mechanisms involving a redundant array of independent disks can help to retain availability.

Several existing research papers published during the window period of 2015–2019 have been reviewed and analyzed based on their effect on security principles, trust and privacy. A reviewed research contribution on Blockchain-based privacy mechanisms for IoT is presented in Table 3.2.

Several abbreviations are used in Table 3.2 and Table 3.3, which are defined in Table 3.1.

3.3 SECURITY CHALLENGES IN IOT

A few security challenges and attacks have been identified in IoT and Blockchained IoT models. Some of the major challenges are discussed in Table 3.3, which outlines the comparative study of specific security risks, their consequences, extent of attack, design paradigm events, breached security principles and their counteracting actions.

3.3.1 Challenges to Blockchain IoT

- **Scalability and Storage:** Scalability can be seen as an essential feature of the process of integration of IoT and Blockchain. As the size of the IoT network increases, the centralized repository of Blockchain may also increase,

TABLE 3.1
Abbreviations

P	Perception Layer
NT	Network and Transmission Layer
AM	Application and Middleware Layer
C	Confidentiality
I	Integrity
A	Availability

TABLE 3.2
Reviewed Research Contributions on Blockchain-based Privacy Mechanisms for IoT

	Authors	Description	Security Principle Affected			Trust	Privacy
			C	I	A		
R1:	(Aitzhan & Svetinovic, 2016)	Using group signatures and anonymous communication sources that are authenticated off-chain to ensure anonymity in energy trading applications. Using the techniques of Blockchain technology, i.e., distributed/decentralization and flexibility in energy smart grids, an encrypted off-chain communication network is established from users to manufacturers for participating in Blockchain agreements.	√	√	√	√	√
R2:	(Dorri et al., 2017)	A multi-layer Blockchain infrastructure is used to exchange encrypted data with all the participants. This is to ensure the throughput management ratio of assigned values (Assigned Throughput Values) with total accessible throughput.	√	√	√	√	√
R3:	(Cha et al., 2018)	Blockchain-linked gateways for handling existing IoT systems and transacting data over the network. The Blockchain interface retains a sense of anonymity, while the network holds immutable encrypted consumer preferences information. So, the gateway improves protection with Bluetooth low-energy devices on the IoT edge.	√	×	√	×	√
R4:	(Ouaddah et al., 2016)	Provides solutions where IoT owners have complete influence over how they want to offer their IoT data exposure with tokenized approach. Enables private IoT data ownership and offers a mechanism under which IoT owners retain complete influence over how they want to allow access to their IoT data.	√	√	√	√	√
R5:	(Zyskind et al., 2015)	The proposed models offer applications for database management, which focuses on Decentralized Hash Tables (DHT) and information received from the decentralized Blockchain blocks. This model finetunes the access control policies, which restricts the users who can manipulate data.	√	√	√	√	√

(*Continued*)

TABLE 3.2 (CONTINUED)
Reviewed Research Contributions on Blockchain-based Privacy Mechanisms for IoT

	Authors	Description	Security Principle Affected			Trust	Privacy
			C	I	A		
R6:	(Shafagh et al., 2017)	Data related to the Blockchain Data Access Management System held in Decentralized Hash Tables (DHT) off-chain. Blockchain reserves authorization rights for diverse users on any data contained in the DHT with tokenized off-chain control rights.	√	√	√	√	√
R7:	(Hardjono & Smith, 2016)	Provides privacy protection when procuring IoT devices across the cloud utilizing certified Blockchains. Also, grants provenance of a resource-restricted IoT system without declaration of its origin.	√	×	√	√	√
R8:	(Ali et al., 2017)	This approach utilizes Inter Planetary File System (IPFS) as a decentralized IoT data storage platform and utilizes multi-layered Blockchain design focused on smart contract-driven access management principles as well as peer-to-peer encryption.	√	√	√	√	√
R9:	(Liu et al., 2017)	Using the Blockchain-based transparent data integrity authentication platform by utilizing smart contracts helps various applications to check the rationality of the data stockpiled at the platforms supporting cloud applications.	√	√	√	√	×
R10:	(Bahga & Madisetti, 2016)	There are some tasks in the network that need to be accomplished upon the IoT devices or nodes as sovereign solicitations to comprehend the upright authentications. This objective can be achieved by coupling the Blockchain technology with industrial IoT.	√	√	√	√	×
R11:	(Boudguiga et al., 2017)	In a distributed framework, for sending the updates, the Blockchain is used to monitor software update transactions that are sent to devices for avoiding fraudulent security updates on nodes. In this situation, there is no need for a trustworthy broker to provide updates, because notifications that are propagated to the users via the Blockchain maintain transparency.	√	√	√	√	×

(Continued)

TABLE 3.2 (CONTINUED)
Reviewed Research Contributions on Blockchain-based Privacy Mechanisms for IoT

	Authors	Description	Security Principle Affected			Trust	Privacy
			C	I	A		
R12:	(Axon & Goldsmith, 2016)	To control IoT apps, this leverages a Blockchain-based Public-Key Infrastructure (PKI). They used smart contracts-based Blockchain addresses, which provided commands to the IoT devices. Such commands vary from modifying operating practices to storing data on energy usage on the Blockchain.	√	√	√	√	×
R13:	(Biswas & Muthukkumarasamy, 2016)	The research presents a model of a smart city based on a Blockchain technology for maintaining the discretion and concealment of scrambled information by the fundamental immutableness of the Blockchain. The suggested approach uses the Blockchain based on Ethereum with smart contracts for the determination of functionality.	√	√	√	√	×
R14:	(Yang et al., 2017)	A scaffold platform for the vehicle based on IoT technology and focused on reliability assessment. The suggested approach consists of a legitimacy network built on Blockchain, which determines the authenticity of the sent messages depending on the sender's identity.	√	√	√	√	×
R15:	(Lee & Lee, 2017)	This entails creating and managing Blockchain identities as a service via Blockchain-based identity and authentication management framework for smartphone applications and IoT nodes without any concern for connections or contact.	√	√	√	√	×
R16:	(Steger et al., 2018)	Evidence for safe dissemination of application upgrades in vehicles based on smart technologies, utilizing a layered Blockchain scalability design.	√	√	√	√	√

(Continued)

TABLE 3.2 (CONTINUED)
Reviewed Research Contributions on Blockchain-based Privacy Mechanisms for IoT

	Authors	Description	Security Principle Affected			Trust	Privacy
			C	I	A		
R17:	(Alphand et al., 2018)	Solution developed for protection and monitoring using IoT technologies and Blockchain to carry out the authorization policies and maintain the interaction logs. This tool also ensures the authentication paradigm of Object Security Architecture for IoT by utilizing a group key scheme.	√	√	√	√	×
R18:	(Chakraborty et al., 2018)	A multi-layered Blockchain approach for addressing resource-limited IoT application protection issues. The communication of lower-layer devices happens using the higher-layer nodes. These inherent properties ensure the viability of the provided solution.	√	√	√	√	×
R19:	(Dwivedi et al., 2019)	The execution of a novel amalgam elucidation blends the benefits of private key technique, public key technique, Blockchain and several other cryptographic primitives.	√	√	√	√	√
R20:	(Dorri et al, 2019)	The researchers have implemented an approach that was based on distributed trust mechanism and decreases the Online Black-Markets' (OBM) processing time for validating new blocks. The technique is also based on an IoT-friendly consensus algorithm, which circumvents the requirement to As an added bonus, the technique is built on an Internet of Things-friendly consensus process that eliminates the need to unravel any delinquent blocks until a block is appended to the Blockchain until a block is appended to the BC.	√	√	√	√	√

TABLE 3.3
Counteracting Actions on Common Security Challenges and Their Effect on Layers and Security Principles

Challenges	Description	Layer Affected			Security Principle Affected			Security Counteract Actions
		P	NT	AM	C	I	A	
Physical Damage by Unauthorized Access	Active attackers may have unauthorized connections to physical sensing instruments, their control systems or their IoT network activities, and owing to lack of technological expertise, attackers can only impact IoT resources by tampering with the physical devices that relate to the network.	√	×	×	√	√	√	Self-destruction, physical design security, tamper proofing (Mosenia & Jha, 2016), hardware-based Trusted Platform Module (TPM) modules, firmware access to USB prevention (OWASP, 2016)
Physical Node Capture	Instead of causing physical harm, an aggressive intruder could seek to access information that the device may add to the network. Instead of directly influencing items, the individual may also target data storage or processing units that ultimately affect the network.	√	×	×	√	√	√	Minimizing leakage by shielding, adding noise, random delays, hamming weights, improving cache architecture (Nia et al., 2015), integration of PUF (physically unclonable function) into objects (Wachsmann & Sadeghi, 2014)
Tag Cloning	Tags are usually mounted on many items and are distributed in open access environments, which presents issues like replication of data, sniffing, authentication and authorization.	√	×	×	√	√	√	Tag seclusion, aloofness approximation, tag hindering, instigating authentication procedures, hash-based arrangements, encryption techniques (Khedr, 2013), One Time Password (OTP) harmonization between tag and back end
Privacy Leak	Data privacy is one of the main concerns, since a broad variety of IoT devices are engaged in data harnessing and transferring it to other network nodes, thus collecting, processing and disposing of data that is no longer required.	√	×	×	√	√	×	Homomorphic encryption, digital signatures, fragmentation redundancy scattering (Smart & Vercauteren, 2010)

(Continued)

TABLE 3.3 (CONTINUED)
Counteracting Actions on Common Security Challenges and Their Effect on Layers and Security Principles

Challenges	Description	Layer Affected			Security Principle Affected			Security Counteract Actions
		P	NT	AM	C	I	A	
Brute Force Attack	An attack that uses automated software to find the appropriate encryption key to bring all probable blends of key sets from the key puddle. However, with the rise in computational capacity and effective rate increases to retrieve the keys using brute force techniques, several other attempts have quickly started to reduce the search space and recover keys.	√	×	×	√	√	√	Cryptography techniques, firmware security update
Eavesdropping	An incursion involving the detection of correspondence between two network points by snuffling or capturing data packets and therefore stealing information between them utilizing cryptographic analysis methods.	×	√	×	√	√	×	Authentication and authorization mechanisms, route security using the secured channel technique, encryption of RFID communication channel (Mitrokotsa et al., 2010), installation of network key over the devices before initialization of process (Community, n.d.), mutual authentication of nodes over the network and data tamper detection mechanism over the nodes residing on the network
Routing Attack	Routing Protocol for Low Power and Lossy Networks, usually denoted as RPL, strongly monitors an intention-positioned regional anatomy of Destination Oriented Directed Acyclic Graph (DODAG). Every node in the network encompasses its corresponding one-to-one unique IDs and segregates these IDs into Parent Node IDs and Neighbor Node IDs.	×	√	×	√	√	√	The security actions may include the Distributed Hash Tables (DHT), warehousing and tracing individualities of each instance over the RPL, integration of signature-based authentication and authorization mechanisms (Dvir et al., 2011; Le et al., 2013; Pongle & Chavan, 2015)

(Continued)

TABLE 3.3 (CONTINUED)
Counteracting Actions on Common Security Challenges and Their Effect on Layers and Security Principles

		Layer Affected			Security Principle Affected			
Challenges	**Description**	**P**	**NT**	**AM**	**C**	**I**	**A**	**Security Counteract Actions**
Flooding Attack	Flooding is a mechanism based on spreading the initial messages over the network internally or externally. It is accountable for maximizing the amount of traffic in the network to make services inaccessible to those who consume them. In the worst cases, this could be by way of accessing certain machines or connections. One of the big attacks in this group is the "Hello" flooding attack.	√	×	×	√	√	√	The security actions may include the computation of the signal strength over the network, packet delivery ratio, encoding of those packets with misplayed amendment codes, changing the incidence and whereabouts, implementation of the firewalls and abysmal packet interception (Noubir & Lin, 2003; What is a UDP Flood DDoS Attack Glossary, n.d.; Xu et al., 2005, 2004)
Wormhole Attack	This influences both the direction of network traffic and the topology of a network. The attack starts with the transmission of all the available packets and establishing a tunnel between two nodes over two different networks.	×	√	×	√	√	×	Conviction administration, supervision of the keys, measurement of the strength of the signal, topographical data fastening and graph traversal (Ahmed & Ko, 2016; Hu et al., 2005; Jang et al., 2007; Krontiris et al., 2007; Lazos et al., 2005; Ngai et al., 2006; Pirzada & McDonald, 2005; Poovendran & Lazos, 2007; Raju & Parwekar, 2016; Salehi et al., 2013; Sharmila & Umamaheswari, 2011; Wang et al., 2008; Wazid et al., 2016; Weekly & Pister, 2012)

(Continued)

TABLE 3.3 (CONTINUED)
Counteracting Actions on Common Security Challenges and Their Effect on Layers and Security Principles

Challenges	Description	Layer Affected			Security Principle Affected			Security Counteract Actions
		P	NT	AM	C	I	A	
Replay Attack	This attack is based on the mechanism whereby a malicious node starts professing to be an authorized node and senses data communication. The malicious node brings on its transmission or interruption into the network, misleading the receiver nodes.	×	√	×	√	√	×	Timestamping, hash chain-based verification of fragments (Vidgren et al., 2013; Xiao et al., 2007)
Spoofing	The forging of the MAC-Media Access Control (MAC) or Internet Protocol (IP) address of any valid node on the network by any malicious nodes may be considered as IoT spoofing. On controlling the identity of the legitimate node, the malicious node starts pretending to be the legitimate node itself and obtains unauthorized access to the IoT network.	×	√	×	√	√	x	Estimating the channel of propagation, signal potency measurement, encryption techniques, authentication of message, and Secure Socket Layer (SSL) and message filtering (Tay et al., 2016)
Sybil Attack	The IoT network may encounter a Sybil attack which enables the malicious nodes to focus their characteristics and propagate junk mail phishing over the network.	×	√	×	√	√	√	Classification-based Sybil detection (BCSD), user behaviour analysis, trusted and untrusted user list maintenance, random walk on social graphs (Alvisi et al., 2013; Cao & Yang, 2013; Mohaisen et al., 2011; Wang et al., 2012; Zhang et al., 2014); Douceur's approach – Trusted certification

(Continued)

TABLE 3.3 (CONTINUED)

Counteracting Actions on Common Security Challenges and Their Effect on Layers and Security Principles

Challenges	Description	Layer Affected			Security Principle Affected			Security Counteract Actions
		P	NT	AM	C	I	A	
Sinkhole Attack	IoT networks that operate mostly in aggressive environments are typically left unattended with devices that have minimal battery power, computing capacities and contactability. These are typically more vulnerable to sinkhole attack, in which a node in the network is breached by an invader, and depending on the routing metrics, it can continue to create contact with adjacent nodes.	×	√	×	√	√	√	Parent fail-over, Intrusion Detection System (IDS) solution, generating identity certificates and rank authentication techniques (Raza et al., 2013; Weekly & Pister, 2012), message digest algorithm (Kibirige & Sanga, 2015)
Denial of Service	Detecting an aggressive atmosphere with battery-operated equipment such as wireless sensor networks renders the system more susceptible to events such as battery leakage, since battery change in these conditions is not feasible due to the lack of infrastructure.	×	×	×	×	×	√	Suspicious devices list maintenance, access control lists, policies provided by providers (Chandna et al., 2014; Haataja, 2005; Ongtang et al., 2012), load balancing (Gupta et al., n.d.)
Sleep Deprivation Attack	Sleep deprivation is one of the most destructive attacks of this nature, where challengers attempt to increase the power usage of nodes in the network to reduce the nodes' wakeup time.	×	√	×	√	√	√	Content chaining approach, multi-layer-based intrusion detection, split buffer approach, target IPv6 defence movement in 6LoWPAN (Bhattasali & Chaki, 2011; Hummen et al., 2013; Sherburne et al., 2014), Random vote, Round Robin scheme (Pirretti et al., 2006)

(*Continued*)

TABLE 3.3 (CONTINUED)
Counteracting Actions on Common Security Challenges and Their Effect on Layers and Security Principles

		Layer Affected			Security Principle Affected			
Challenges	**Description**	**P**	**NT**	**AM**	**C**	**I**	**A**	**Security Counteract Actions**
Insecure Software, Firmware and Interfaces	The applications used for accessing IoT resources are focused on cloud, web or smartphone applications that are extremely vulnerable to attacks and may therefore impact the privacy of data. In comparison to interfaces, vulnerabilities may be triggered by insecure hardware or applications, and so their upgrades must be carried out safely.	×	×	√	√	√	×	Regular device updates, file encryption using acceptable encryption techniques, file transmission via encrypted connection, secured update server (OWASP, 2016)
CoAP and Middleware Security	CoAP implements a message format specified in RFC-7252 to provide end-to-end protection in restricted applications and uses Datagram Transport Layer Security (DTLS) connections with several security modes. CoAP messages based on RFC-7252 require encryption for safe communication because CoAP multicast support needs authentication and key management.	×	×	√	√	×	×	VIRTUS Middleware(Conzon et al., 2012), security policies, Secure Middleware for Embedded Peer-to-Peer systems (SMEPP) (Caro et al., 2009), lightweight DTLS (Rescorla & Modadugu, 2012), Transport Layer Security- Datagram Transport Layer Security (TLS-DTLS) mapping, Hyper Text Trasfer Protocol (HTTP)-CoAP mapping, TLS-DTLS tunnelling, message filtration using 6LBR, service layer Machine to Machine (M2M) security (Brachmann, Keoh, et al., 2012; Granjal et al., 2013; Sethi et al., 2012; Brachmann, Garcia-Mochon, et al., 2012; Caro et al., 2009; Conzon et al., 2012; Ferreira et al., 2014; Gómez-Goiri et al., 2014; OneM2M, Security Solutions – OneM2M Technical Specification, 2017; Rescorla & Modadugu, 2012)

Source: Gupta, M., et al., Security issues in Internet of Things: Principles, challenges, taxonomy. In Proceedings of Springer Lecture Notes Electrical Engineering at 3rd International Conference on Recent Innovations in Computing (ICRIC-2020), 2020.

which further increases transaction numbers. With the scalable architecture and increasing number of transactions of the Blockchain ledger, the need for more storage comes into the picture, which can be met using cloud storage solutions in the next phase, where various backups and replications needed will make the whole system more complex.

- **Processing Power and Time:** There are several encryption algorithms used in integrating Blockchain with IoT technology. The ecosystem of the IoT provides a different paradigm for the computing capabilities and security algorithms contained in it, requiring more and more processing power.
- **Shortage of Experienced and Skilled Developers:** One of the most prominent issues in integrated technology is the shortage of skilled employees. The two technologies are poles apart, and the integration requires personnel with the skills to develop and handle this.
- **Policy Management, Legal Guidance and Compliance:** The policy management and legal compliance are again an important concern in the case of the Blockchained IoT ecosystem. The policies related to privacy of the data may vary from place to place, so the algorithms used in the integration should have common derivatives.
- **Complexities of Hybrid and Integrated Structure:** As discussed, both the fields are emerging and complex in nature, so the amalgamation of both technologies will give rise to more complex systems. The current IoT ecosystems bank on the centralized system and client server communication models, whereas Blockchain is a decentralized technology. The resource constraints and infrastructural requirements will make the system difficult to implement and handle.
- **Security Issues:** Current IoT systems do not offer multi-factor authentication. The market for Blockchain as a service is very new, and the processing power needs can be an issue in physical systems, so the whole system is dependent on the cloud. So, vendor-level risks are associated with the technology, which needs to be addressed. The IoT components must be durably connected to the block network to participate in the process generated by Blockchain to ensure reliability.

Besides these basic challenges, some of the major attacks on the Blockchain described here with their analysis in various reviewed studies of Blockchained IoT, mentioned in Table 3.2, are described in brief in Table 3.4.

- **Denial of Service (DoS) Attack**

  The invader attempts to prevent the authentic user from accessing the service in the network during such an attack. In these cases, the adversary may start fraudulent transactions and may increase network traffic. Random users cannot access the network in our program, however, without evidence of authority. Nevertheless, assume that an adversary invades the network and begins to send fake transactions to the network. For such a scenario, the cluster head will search the public address, and if it is not available or

TABLE 3.4
Coverage and Analysis in Reviewed Research

Attack	Denial of Service (DoS) Attack	Distributed DOS (DDoS) attack	51% Attack	Double-spending	Mining Attack	Collision Attack	Linking Attack	Modification Attack	Dropping Attack	Sybil Attack
R1	N	Y	Y	Y	N	Y	X	N	N	Y
R2	N	Y	N	N	N	N	Y	N	N	N
R3	X	N	N	N	N	N	N	N	N	N
R4	N	N	N	N	N	N	N	N	N	N
R5	N	N	N	N	N	N	N	N	N	N
R6	N	N	N	N	N	N	N	N	N	N
R7	N	N	N	N	N	N	N	N	N	Y
R8	N	N	N	N	N	N	N	N	N	N
R9	N	N	N	Y	N	N	N	N	N	N
R10	N	N	N	N	N	N	N	N	N	N
R11	Y	N	N	N	N	N	N	N	N	N
R12	N	N	N	N	N	N	N	N	N	N
R13	N	N	N	N	N	N	N	N	N	N
R14	N	N	N	N	N	N	N	N	N	N
R15	N	N	N	N	N	N	N	N	N	N
R16	N	N	N	N	N	N	N	N	N	N
R17	Y	N	Y	N	N	N	N	N	N	Y
R18	N	N	N	N	N	N	N	N	N	N
R19	Y	N	Y	N	Y	N	N	N	Y	N
R20	N	Y	Y	N	N	N	Y	Y	Y	N

registered with the cluster head, then the transaction will not be transferred in the network and accelerated to other clusters.

- **Distributed DOS (DDoS) Attack**

  DDoS attacks render certain facilities inaccessible by overwhelming them with unnecessary counterfeit traffic. The traffic uses space and energy, such that the platform is shut down. Mainly due to a lack of adequate security features for connected IoT devices, the potential for disruption and the choice of manufacturing costs and profit potentials have been under-estimated in the past. However, recent events have triggered a rethink to include and promote safety characteristics for connected products as part of a premium positioning. Now, it may not be simple for a professional intruder to manipulate IoT devices and build a botnet.
- **51% Attack**

  A specific weakness in Blockchain implementations is the 51% attack, where the hash rate of the network bulk for accessing the database is obtained by one intruder, a community of Sybil nodes or a mining pool on the network. Attackers would avoid confirmation of new transactions, which would allow them to stop transactions between some or all users. In rare circumstances, attackers with a hash rate of over 50% can take other miners and add their blocks with a high likelihood to the Blockchain; in addition, fraudulent or double-spent transactions may occur.
- **Double-Spending**

  Whether twice or on many occasions, double-spending involves the usage of a onetime transaction. A transaction moves asset ownership from the identity of a sender to the public address of the receiver, and the signatory is required to sign the transaction using a private key. After the transaction has been signed, it will be transmitted to the network on which the transaction is validated. The receiver searches for the sender's unexpended transaction, searches the signature of the sender and assumes that a transaction is mined in a legitimate chain.
- **Mining Attack**

  In this attack, a few cluster heads are hacked by an adversary, who starts manipulating several of them; false mining is possible in such a situation, but if other cluster heads or nodes detect it, they can easily track the false cluster heads. If the network senses a false cluster head, it can be altered by the nodes in that cluster.
- **Collision Attack**

  A collision attack on an authenticated hash attempts to trigger a hash collision, that is, two inputs of the same hash value. If an attacker succeeds, this could generate two public keys with the same address. The intruder does not monitor the content of any message during a classical collision attack; it is randomly chosen by the algorithm.
- **Linking Attack**

  An attacker, who could be a cloud storage or service provider, connects several cloud data or Blockchain transactions with the same ID to establish an anonymous node's real-world identity. Any information relating to private data of the user can be identified by an attacker; an intruder may attempt de-anonymization of users by linking a user's PKs (Primary Keys) and ties to separate data pieces with the same anonymous user for a period.

- **Modification Attack**

  The attacker must compromise the integrity of cloud storage to initiate this attack. The attacker may then try to modify or delete stored information for a certain user. This user can identify any improvements in the storage data by matching the cloud hash with the local BC (Blockchain) hash. When a user identifies a privacy infringement, he produces a transaction implying two transactions: The user-signed transaction and the cloud storage, which include the real method of the privacy and the user-signed access transaction containing the database and the user-invalid hash of the data. The transaction is then forwarded to many CHs (Cluster Heads) that verify the initial transactions. In the event of incoherence of two hashes, CH alerts the cloud storage nodes to malicious activities. However, when subjected to this attack, the user cannot recover his data.

- **Dropping Attack**

  The adversary will monitor a CH or group of CHs to launch this attack. The CHs managed by the attacker will delete all transactions and blocks that have been issued. Nonetheless, an intrusion will be observed because of the lack of transactions or resources from the network from any nodes belonging to the constituent clusters.

- **Sybil Attack**

  A Sybil attack is a security vulnerability on an electronic infrastructure that aims to take over the network by the creation of several identities, nodes and machines. Attackers will deny the authentic nodes over the network if appropriate false identities are generated and then fail to accept or send blocks and obstruct other network users. Most Blockchains employ different "algorithms of consensus" to better protect themselves against Sybil attacks. These contain work proof, proof of stake and vicarious proof of stake. Such algorithms of consensus do not eliminate attacks by Sybil, but they render an effective attack by Sybil very impractical for an intruder.

3.4 APPLICATIONS OF BLOCKCHAINED IOT

There are numerous applications of Blockchain and IoT. The amalgamation of both the technologies has come up with new possibilities. The following case studies can be seen as the platforms where the amalgamated models can be used.

3.4.1 Case Study I: Smart Home Based on Blockchained IoT

A Blockchain-based smart home application is an example where this combined technology can be seen. In IoT, security and privacy are the major concerns due to the wide-ranging nature of IoT devices (Dorri et al., 2017). A large number of researchers are working on it. A smart home application can be the integration of three technologies together; cloud technology for storage, Blockchain ledger for maintaining the records of every transaction, and IoT devices to develop a smart home. Miners are used in smart homes to enable all high-resource devices to communicate with each other internally and externally. Blockchain is used in smart homes for controlling and auditing the communication channel. The process starts with the initialization of the IoT devices, processing of transactions, communication between the devices, working of miners, security, etc. Communication between the local devices and the network is known as a transaction, and according to the security point, access and requirements, different transactions such as access transaction (to use cloud storage), monitor transaction (to keep an eye on devices), genesis transaction (to add device) and remove transaction (to remove devices) are used in smart homes. A shared key is used between all transactions to perform secure communication, which is distributed by the miner. Blockchain, which is used locally in the network, can maintain the record of all transactions according to the policy related to that particular transaction. The miner is responsible for processing all incoming and outgoing transactions from the smart home. All transaction history is stored in Blockchain, and an external storage device is attached to the miner to create backup of the data. All the incoming and outgoing messages will be authorized by the miner to protect the smart home from attacks. Hierarchical design, for which different levels are embedded in the system, can be used to protect the system from DDoS attacks. To secure a smart home, IPv6 over Low Power Wireless Personal Area Networks (6LoWPAN) was used for simulation with three z1 mote sensors sending data to the miner every 10 seconds. It was concluded that the overheads of the proposed system were low and manageable for low-resource IoT devices. The Australia-based telecommunication and media company "Telsetra" provides smart home solutions and has implemented a Blockchain-based solution to ensure data integrity in smart homes.

3.4.2 Case Study II: Supply Chain Management Solution Using Blockchained IoT

Supply chain management (SCM) refers to the large range of business activities that require planning, controlling, tracking and execution of product flows from obtaining raw materials to production and distribution to the end users or consumers. Every trader (merchant) and manager of a firm wants to have access to the processed data all the time. Along with the accessibility of the data, security of data and integrity are the important aspects when business is concerned. And of course, the other important aspect is the health of machinery, which is the backbone of production companies. From the start, the business world has always wanted to spend less and gain more profit, along with achieving more customers. There is a lack of transparency in the current supply chain models used in logistics. A combined model can be helpful in attaining the traceability of packages in the entire network. There is a need for a higher security level where labelled data processing can generate automated settlement responses. The secure delivery of packages is a mandatory condition in shipping logistic processes. The inclusion of Blockchain in IoT-based ecosystems to be adopted in SCM will create a system that will be immutable, traceable and automated. Pooled with crypto-empowered hardware, Blockchain significantly intensifies security at all the levels of the SCM process. Examples of this system can be seen at Golden State Foods (GSF), where a diversified model along with IBM is developed to optimize the business process.

3.4.3 Case Study III: Blockchain-based IoT Solution for Automotive Transportation

The automotive transportation industry is not untouched by security issues. The use of various sensors in the vehicle is making them capable of driving themselves without any driving support. There is a problem with the design and commercialization of such vehicles, as security is the main concern. If a vehicle is hacked and commits a crime, how can the law deal with this situation? The use case of Blockchain comes here with a solution where the inculcation of it will make the automotive technology more reliable. The Blockchain and IoT can help to make the technology more convenient to use by initiating automatic fuel payments, smart parking systems and automatic traffic control systems. A smart parking system has been demonstrated using the Blockchain and IoT coupled model. The integrated model helps to find vacant space in parking areas and generates automatic payments using digital wallets.

3.4.4 Case Study IV: Blockchain-based IoT Solution for Agriculture

Agriculture cannot be considered an industry, as one can live without luxury but cannot live without food. There is a lot of potential in the food production industry. The installation of various types of IoT sensors in farms and moving the sensed data directly to the Blockchained ledger can enrich the process of food supply logistics. PAVO, one of the players in smart agriculture solution providers, has used an integrated approach that brings transparency. The company provides a hardware-based solution whereby IoT devices are installed in farms and communicate the data directly to the Blockchained ledger. This solution enables users with new farming techniques by capturing the data and informing the stakeholders of the decisions.

3.5 CONCLUSION AND FUTURE SCOPE

Although Blockchain still has several hurdles to conquer, as the future of Blockchain, applications in IoT are extremely bright. Internation Data Corporation (IDC) predicts that by 2025, as many as 45% of IoT implementations will use Blockchain technologies. There are several interesting ways to utilize Blockchain to increase efficiency across the supply chain. But, once we get there, we need to ask some basic questions about long life, security and application. So, we need answers to guarantee that the "20% of operations" are as effective, productive and stable as possible. The technology of Blockchain is in the early stages and heading for a peak of inflated experimentation. To reach the plateau of stability and coming into production, it needs to go through the trough of disillusionment and the slope of enlightenment. This chapter has tried to cover all the attacks that an IoT network can face. All the attacks are discussed with respect to the layered model of IoT framework and the important aspects of security, i.e., Confidentiality, Integrity and Availability. The amalgamation of Blockchain technology with IoT-enabled devices has achieved new dimensions to the security frameworks, but there is still scope for improvement in this segment of technology. In the coming 10 years, this technology will be booming and integrated into smart homes, smart cities, smart offices and smart devices as a security framework.

REFERENCES

Ahmed, F., & Ko, Y.-B. (2016). Mitigation of black hole attacks in routing protocol for low power and lossy networks. *Security and Communication Networks*, *9*(18), 5143–5154.

Ahram, T., Sargolzaei, A., Sargolzaei, S., Daniels, J., & Amaba, B. (2017). Blockchain technology innovations. In 2017 IEEE Technology & Engineering Management Conference (TEMSCON). Santa Clara, USA (pp. 137–141).

Aitzhan, N.Z., & Svetinovic, D. (2016). Security and privacy in decentralized energy trading through multi-signatures, Blockchain and anonymous messaging streams. *IEEE Transactions on Dependable and Secure Computing*, *15*(5), 840–852.

Ali, M.S., Dolui, K., & Antonelli, F. (2017). IoT data privacy via Blockchains and IPFS. In Proceedings of the Seventh International Conference on the Internet of Things - Linz Austria (pp. 1–7).

Alphand, O., Amoretti, M., Claeys, T., Dall'Asta, S., Duda, A., Ferrari, G., Rousseau, F., Tourancheau, B., Veltri, L., & Zanichelli, F. (2018). IoTChain: A Blockchain security architecture for the Internet of Things. In 2018 IEEE Wireless Communications and Networking Conference (WCNC) – Barcelona Spain (pp. 1–6).

Alvisi, L., Clement, A., Epasto, A., Lattanzi, S., & Panconesi, A. (2013). Sok: The evolution of sybil defense via social networks. In 2013 IEEE Symposium on Security and Privacy (pp. 382–396).

Axon, L.M., & Goldsmith, M. (2016). *PB-PKI: A privacy-aware Blockchain-based PKI.* Oxford

Bahga, A., & Madisetti, V.K. (2016). Blockchain platform for industrial internet of things. *Journal of Software Engineering and Applications*, *9*(10), 533–546.

Bhattasali, T., & Chaki, R. (2011). A survey of recent intrusion detection systems for wireless sensor network. In International Conference on Network Security and Applications (pp. 268–280).

Biswas, K., & Muthukkumarasamy, V. (2016). Securing smart cities using Blockchain technology. In 2016 IEEE 18th International Conference on High Performance Computing and Communications; IEEE 14th International Conference on Smart City; IEEE 2nd International Conference on Data Science and Systems (HPCC/SmartCity/DSS) Sydney, Australia (pp. 1392–1393).

Boudguiga, A., Bouzerna, N., Granboulan, L., Olivereau, A., Quesnel, F., Roger, A., & Sirdey, R. (2017). Towards better availability and accountability for iot updates by means of a Blockchain. In 2017 IEEE European Symposium on Security and Privacy Workshops (EuroS&PW) (pp. 50–58).

Brachmann, M., Garcia-Mochon, O., Keoh, S.-L., & Kumar, S.S. (2012a). Security considerations around end-to-end security in the IP-based Internet of things. In Workshop on Smart Object Security, in Conjunction with IETF83, Paris, France, March 23, 2012.

Brachmann, M., Keoh, S.L., Morchon, O.G., & Kumar, S.S. (2012b). End-to-end transport security in the IP-based internet of things. In 2012 21st International Conference on Computer Communications and Networks (ICCCN) Munich, Germany (pp. 1–5).

Cao, Q., & Yang, X. (2013). SybilFence: Improving social-graph-based sybil defenses with user negative feedback. ArXiv:1304.3819.

Caro, R.J., Garrido, D., Plaza, P., Roman, R., Sanz, N., & Serrano, J.L. (2009). Smepp: A secure middleware for embedded p2p. In ICT Mobile and Wireless Communications Summit (ICT-MobileSummit'09).

Cha, S.-C., Chen, J.-F., Su, C., & Yeh, K.-H. (2018). A Blockchain connected gateway for BLE-based devices in the Internet of Things. *IEEE Access, 6*, 24639–24649.

Chakraborty, R.B., Pandey, M., & Rautaray, S.S. (2018). Managing computation load on a Blockchain-based multi-layered Internet-of-Things network. *Procedia Computer Science, 132*, 469–476.

Chandna, S., Singh, R., & Akhtar, F. (2014). Data scavenging threat in cloud computing. *International Journal of Advances In Computer Science and Cloud Computing, 2*(2), 106–111.

Community, T.C.S. (n.d.). *Cyber Security Community. Different Attacks and Counter Measures Against ZigBee Networks.* Retrieved July 20, 2020, from https://securitycommunity.tcs.com/

Conzon, D., Bolognesi, T., Brizzi, P., Lotito, A., Tomasi, R., & Spirito, M.A. (2012). The virtus middleware: An xmpp based architecture for secure iot communications. In 2012 21st International Conference on Computer Communications and Networks (ICCCN) (pp. 1–6).

Dorri, A., Kanhere, S S., Jurdak, R., & Gauravaram, P. (2017). Blockchain for IoT security and privacy: The case study of a smart home. In 2017 IEEE International Conference on Pervasive Computing and Communications Workshops (PerCom Workshops) (pp. 618–623).

Dorri, A., Kanhere, S.S., Jurdak, R., & Gauravaram, P. (2019). LSB: A lightweight scalable Blockchain for IoT security and anonymity. *Journal of Parallel and Distributed Computing, 134*, 180–197.

Dvir, A., Buttyan, L., & others. (2011). VeRA-version number and rank authentication in RPL. In 2011 IEEE Eighth International Conference on Mobile Ad-Hoc and Sensor Systems Valencia, Spain (pp. 709–714).

Dwivedi, A.D., Srivastava, G., Dhar, S., & Singh, R. (2019). A decentralized privacy-preserving healthcare Blockchain for IoT. *Sensors, 19*(2), 326–332.

Ferreira, H.G.C., de Sousa, R.T., de Deus, F.E.G., & Canedo, E.D. (2014). Proposal of a secure, deployable and transparent middleware for internet of things. In 2014 9th Iberian Conference on Information Systems and Technologies (CISTI) Barcelona, Spain (pp. 1–4).

Gartner. (2017). *Top Trends in the Gartner Hype Cycle for Emerging Technologies.* http://www.gartner.com/smarterwithgartner/top-trends-in-the-gartner-hype-cycle-for-emerging-technologies-2017/

Gómez-Goiri, A., Orduña, P., Diego, J., & López-De-Ipiña, D. (2014). Otsopack: Lightweight semantic framework for interoperable ambient intelligence applications. *Computers in Human Behavior, 30*, 460–467.

Granjal, J., Monteiro, E., & Silva, J.S. (2013). Application-layer security for the WoT: Extending CoAP to support end-to-end message security for Internet-integrated sensing applications. In International Conference on Wired/Wireless Internet Communication Russia (pp. 140–153).

Gupta, M., Jain, S., Patel, R.B., (2020). Security issues in Internet of Things: Principles, challenges, taxonomy. In Proceedings of Springer Lecture Notes Electrical Engineering at 3rd International Conference on Recent Innovations in Computing (ICRIC-2020) India, In Press.

Gupta, M., Gopalakrishnan, G., & Sharman, R. (2017). Countermeasures against distributed denial of service. School of Management State University of New York Buffalo, NY.

Gupta, S., & Vyas, S. (2021). IoT in Green Engineering Transformation for Smart Cities. In *Smart IoT for Research and Industry* (pp. 121–131). Cham: Springer.

Gupta, S., Vyas, S., & Sharma, K.P. (2020, March). A survey on security for IoT via machine learning. In 2020 International Conference on Computer Science, Engineering and Applications (ICCSEA) India (pp. 1–5). IEEE.

Haataja, K. (2005). Bluetooth network vulnerability to disclosure, integrity and denial-of-service attacks. In Proceedings of the Annual Finnish Data Processing Week at the University of Petrozavodsk (FDPW'2005), Advances in Methods of Modern Information Technology (Vol. 7, pp. 63–103)

Hardjono, T., & Smith, N. (2016). Cloud-based commissioning of constrained devices using permissioned Blockchains. In Proceedings of the 2nd ACM International Workshop on IoT Privacy, Trust, and Security (pp. 29–36).

Hu, Y.-C., Perrig, A., & Johnson, D.B. (2005). Ariadne: A secure on-demand routing protocol for ad hoc networks. *Wireless Networks, 11*(1–2), 21–38.

Hummen, R., Hiller, J., Wirtz, H., Henze, M., Shafagh, H., & Wehrle, K. (2013). 6LoWPAN fragmentation attacks and mitigation mechanisms. In Proceedings of the Sixth ACM Conference on Security and Privacy in Wireless and Mobile Networks USA (pp. 55–66).

Jang, J., Kwon, T., & Song, J. (2007). A time-based key management protocol for wireless sensor networks. In International Conference on Information Security Practice and Experience (pp.314–328).

Khedr, W.I. (2013). SRFID: A hash-based security scheme for low cost RFID systems. *Egyptian Informatics Journal, 14*(1), 89–98.

Kibirige, G.W., & Sanga, C. (2015). A survey on detection of sinkhole attack in wireless sensor network. arXiv preprint arXiv:1505.01941.

Kishore, K., & Sharma, S. (2016). Evolution of wireless sensor networks as the framework of Internet of Things: A review. *International Journal of Engineering Research and Management Technology, 5*, 12.

Kishore, S.S. (2015). A review on energy efficient routing protocols & techniques in wireless sensor networks. *International Journal of Advanced Computer Science & Software Engineering, 5*(9), 552–558.

Krontiris, I., Dimitriou, T., Giannetsos, T., & Mpasoukos, M. (2007). Intrusion detection of sinkhole attacks in wireless sensor networks. In International Symposium on Algorithms and Experiments for Sensor Systems, Wireless Networks and Distributed Robotics (pp. 150–161).

Lazos, L., Poovendran, R., Meadows, C., Syverson, P., & Chang, L. (2005). Preventing wormhole attacks on wireless ad hoc networks: A graph theoretic approach. *IEEE Wireless Communications and Networking Conference*, USA *2*, 1193–1199.

Le, A., Loo, J., Lasebae, A., Vinel, A., Chen, Y., & Chai, M. (2013). The impact of rank attack on network topology of routing protocol for low-power and lossy networks. *IEEE Sensors Journal, 13*(10), 3685–3692.

Lee, B., & Lee, J.-H. (2017). Blockchain-based secure firmware update for embedded devices in an Internet of Things environment. *The Journal of Supercomputing, 73*(3), 1152–1167.

Liu, B., Yu, X.L., Chen, S., Xu, X., & Zhu, L. (2017). Blockchain based data integrity service framework for IoT data. In 2017 IEEE International Conference on Web Services (ICWS) (pp. 468–475).

Maple, C. (2017). Security and privacy in the internet of things. *Journal of Cyber Policy, 2*(2), 155–184.

Mitrokotsa, A., Rieback, M.R., & Tanenbaum, A.S. (2010). Classifying RFID attacks and defenses. *Information Systems Frontiers, 12*(5), 491–505.

Mohaisen, A., Hopper, N., & Kim, Y. (2011). Keep your friends close: Incorporating trust into social network-based sybil defenses. In 2011 Proceedings IEEE INFOCOM China (pp. 1943–1951).

Mosenia, A., & Jha, N.K. (2016). A comprehensive study of security of internet-of-things. *IEEE Transactions on Emerging Topics in Computing*, *5*(4), 586–602.

Nasurudeen, T.F.K., Shukla, V.K., & Gupta, S. (2021, June). Automation of Disaster Recovery and Security in Cloud Computing. In 2021 International Conference on Communication information and Computing Technology (ICCICT) Mumbai, India(pp. 1–6). IEEE.

Ngai, E.C.H., Liu, J., & Lyu, M.R. (2006). On the intruder detection for sinkhole attack in wireless sensor networks. In 2006 IEEE International Conference on Communications Istanbul, Turkey (Vol. 8, pp. 3383–3389).

Nia, A.M., Sur-Kolay, S., Raghunathan, A., & Jha, N.K. (2015). Physiological information leakage: A new frontier in health information security. *IEEE Transactions on Emerging Topics in Computing*, *4*(3), 321–334.

Nouhir, G., & Lin, G. (2003). Low-power DoS attacks in data wireless LANs and countermeasures. *ACM SIGMOBILE Mobile Computing and Communications Review*, *7*(3), 29–30.

OneM2M, Security solutions –OneM2M Technical Specification. (2017). http://onem2m.org/technical/latest-drafts

Ongtang, M., McLaughlin, S., Enck, W., & McDaniel, P. (2012). Semantically rich application-centric security in Android. *Security and Communication Networks*, *5*(6), 658–673.

Ouaddah, A., Abou Elkalam, A., & Ait Ouahman, A. (2016). FairAccess: A new Blockchain-based access control framework for the Internet of Things. *Security and Communication Networks*, *9*(18), 5943–5964.

OWASP. (2016). *Top IoT Vulnerabilities*. https://www.owasp.org/index.php/Top_IoT_Vulnerabilities

Pirretti, M., Zhu, S., Vijaykrishnan, N., McDaniel, P., Kandemir, M., & Brooks, R. (2006). The sleep deprivation attack in sensor networks: Analysis and methods of defense. *International Journal of Distributed Sensor Networks*, *2*(3), 267–287.

Pirzada, A.A., & McDonald, C. (2005). Circumventing sinkholes and wormholes in wireless sensor networks. In IWWAN'05: Proceedings of International Workshop on Wireless Ad-Hoc Networks (Vol. 71).

Pongle, P., & Chavan, G. (2015). A survey: Attacks on RPL and 6LoWPAN in IoT. In 2015 International Conference on Pervasive Computing (ICPC) Pune, India (pp. 1–6).

Poovendran, R., & Lazos, L. (2007). A graph theoretic framework for preventing the wormhole attack in wireless ad hoc networks. *Wireless Networks*, *13*(1), 27–59.

Purohit, R., & Bhargava, D. (2017). An illustration to secured way of data mining using privacy preserving data mining. *Journal of Statistics and Management Systems*, *20*(4), 637–645.

Raju, I., & Parwekar, P. (2016). Detection of sinkhole attack in wireless sensor network. In Proceedings of the Second International Conference on Computer and Communication Technologies India (pp. 629–636).

Raza, S., Wallgren, L., & Voigt, T. (2013). SVELTE: Real-time intrusion detection in the Internet of Things. *Ad Hoc Networks*, *11*(8), 2661–2674.

Rescorla, E., & Modadugu, N. (2012). *Datagram transport layer security version 1.2.*

Salehi, S.A., Razzaque, M.A., Naraei, P., & Farrokhtala, A. (2013). Detection of sinkhole attack in wireless sensor networks. In 2013 IEEE International Conference on Space Science and Communication (IconSpace) India (pp. 361–365).

Sethi, M., Arkko, J., & Keränen, A. (2012). End-to-end security for sleepy smart object networks. In 37th Annual IEEE Conference on Local Computer Networks-Workshops Clearwater USA (pp. 964–972).

Shafagh, H., Burkhalter, L., Hithnawi, A., & Duquennoy, S. (2017). Towards Blockchain-based auditable storage and sharing of iot data. In Proceedings of the 2017 on Cloud Computing Security Workshop (pp. 45–50).

Sharma, K.K., & Sharma, S. (2013). Information security & privacy in real life-threats & mitigations:A review. *IJCST, 4*(3).

Sharma, S., & Kishore, K. (2017). Data dissemination algorithm using cloud services: A proposed integrated architecture using IoT. In 2nd International Conference on Innovative Research in Engineering Science and Technology (IREST-2017) (pp. 7–8), Eternal University, Baru Sahib, Sirmour (HP), India.

Sharma, S., Kishore, K., & India 1&2, S.H.P. (2017). Internet of Things (IoT): A review of integration of precedent, existing & inevitable technologies. *AGU International Journal of Engineering and Technology, 4*, 442–455.

Sharmila, S., & Umamaheswari, G. (2011). Detection of sinkhole attack in wireless sensor networks using message digest algorithms. In 2011 International Conference on Process Automation, Control and Computing USA (pp. 1–6).

Sherburne, M., Marchany, R., & Tront, J. (2014). x (pp. 37–40).

Singh, S., & Singh, N. (2016). Blockchain: Future of financial and cyber security. In 2016 2nd International Conference on Contemporary Computing and Informatics (IC3I) India (pp. 463–467).

Smart, N.P., & Vercauteren, F. (2010). Fully homomorphic encryption with relatively small key and ciphertext sizes. In International Workshop on Public Key Cryptography (pp. 420–443)..

Steger, M., Dorri, A., Kanhere, S.S., Römer, K., Jurdak, R., & Karner, M. (2018). Secure wireless automotive software updates using Blockchains: A proof of concept. In *Advanced Microsystems for Automotive Applications 2017* (pp. 137–149). Cham: Springer.

Suo, H., Wan, J., Zou, C., & Liu, J. (2012). Security in the internet of things: A review. In 2012 International Conference on Computer Science and Electronics Engineering (Vol. 3, pp. 648–651).

Tay, H.J., Tan, J., & Narasimhan, P. (2016). A survey of security vulnerabilities in bluetooth low energy beacons. In *Carnegie Mellon University Parallel Data Lab Technical Report CMU-PDL-16-109.*

Treleaven, P., Brown, R.G., & Yang, D. (2017). Blockchain technology in finance. *Computer, 50*(9), 14–17.

Vidgren, N., Haataja, K., Patino-Andres, J.L., Ramirez-Sanchis, J.J., & Toivanen, P. (2013). Security threats in ZigBee-enabled systems: Vulnerability evaluation, practical experiments, countermeasures, and lessons learned. In 2013 46th Hawaii International Conference on System Sciences ,USA (pp. 5132–5138).

Wachsmann, C., & Sadeghi, A.-R. (2014). Physically unclonable functions (PUFs): Applications, models, and future directions. *Synthesis Lectures on Information Security, Privacy, & Trust, 5*(3), 1–91.

Wang, G., Mohanlal, M., Wilson, C., Wang, X., Metzger, M., Zheng, H., & Zhao, B.Y. (2012). Social turing tests: Crowdsourcing sybil detection. ArXiv:1205.3856.

Wang, W., Kong, J., Bhargava, B., & Gerla, M. (2008). Visualisation of wormholes in underwater sensor networks: A distributed approach. *International Journal of Security and Networks, 3*(1), 10–23.

Wazid, M., Das, A.K., Kumari, S., & Khan, M.K. (2016). Design of sinkhole node detection mechanism for hierarchical wireless sensor networks. *Security and Communication Networks, 9*(17), 4596–4614.

Weekly, K., & Pister, K. (2012). Evaluating sinkhole defense techniques in RPL networks. In 2012 20th IEEE International Conference on Network Protocols Austin, USA (ICNP) (pp. 1–6).

What is a Udp Flood Ddos Attack Glossary. (n.d.). Retrieved July 20, 2020, from https://www.incapsula.com/ddos/attack-glossary/udp-flood.html

Xiao, Q., Boulet, C., & Gibbons, T. (2007). RFID security issues in military supply chains. In The Second International Conference on Availability, Reliability and Security Vienna, Austria (ARES'07) (pp. 599–605).

Xiaohui, X. (2013). Study on security problems and key technologies of the internet of things. In 2013 International Conference on Computational and Information Sciences, China (pp. 407–410).

Xu, W., Wood, T., Trappe, W., & Zhang, Y. (2004). Channel surfing and spatial retreats: Defenses against wireless denial of service. In Proceedings of the 3rd ACM Workshop on Wireless Security (pp. 80–89).

Xu, W., Trappe, W., Zhang, Y., & Wood, T. (2005). The feasibility of launching and detecting jamming attacks in wireless networks. In Proceedings of the 6th ACM International Symposium on Mobile Ad Hoc Networking and Computing (pp. 46–57).

Yang, Z., Zheng, K., Yang, K., & Leung, V.C.M. (2017). A Blockchain-based reputation system for data credibility assessment in vehicular networks. In 2017 IEEE 28th Annual International Symposium on Personal, Indoor, and Mobile Radio Communications, QC, Canada(PIMRC) (pp. 1–5).

Zhang, K., Liang, X., Lu, R., & Shen, X. (2014). Sybil attacks and their defenses in the internet of things. *IEEE Internet of Things Journal, 1*(5), 372–383.

Zhang, L., & Wang, Z. (2006). Integration of RFID into wireless sensor networks: Architectures, opportunities and challenging problems. In 2006 Fifth International Conference on Grid and Cooperative Computing Workshops, Hunan, China (pp. 463–469).

Zyskind, G., Nathan, O., & Pentland, A. (2015). Enigma: Decentralized computation platform with guaranteed privacy. ArXiv:1506.03471.

4 Survey of Blockchain Techniques for IoT Device Security

Rahul Das and Ajay Prasad

CONTENTS

4.1 INTRODUCTION

The Internet of Things (IoT) came into the limelight in September 2003 with the launch of a management vision of a supply chain that can be automatically tracked by Auto-ID (Uckelmann et al. 2011).

With the advancement of technology, smarter and smaller devices have been in demand along with the requirement for automation to reduce human intervention. (Mattern and Floerkemeier 2010) These requirements have accelerated the development of the IoT.

IoT leads to an interconnected world, where devices can be linked to the Internet and controlled remotely. Such devices are referred to as *smart devices*. They can interact with one another, among themselves, with other users or with other IoT devices (Atzori, Iera, and Morabito 2010). IoT devices are mostly built to serve a particular purpose and have limited resources, such as minimal communication and

DOI: 10.1201/9781003138082-4

minimal computational power. Though this minimalistic build may seem to be a challenge, it has helped reduce cost with smaller size and easy deployment where limited network and electricity are available.

4.2 IPv6 AND 5G – ACCELERATING IoT GROWTH

IPv4, which was till recently the addressing standard for the Internet, has limited address space. IPv4 posed a challenge to the IoT, as limited IP addresses meant that only a limited number of devices could be connected to the Internet. The introduction and adoption of IPv6 paved the way for IoT, as it meant that more devices now could be connected to the Internet with an IP address. IPv6 addressed the challenges arising from addressing and reduced the complexity of NAT-ing (Network Address Translation) and other network-related challenges.

The future of the IoT is now being accelerated with the introduction of fifth-generation networks (5G). As per the International Data Corporation (IDC) report, the global 5G facilities will result in 70% of companies investing $1.2 billion on connection management solutions ("5G and IoT: The Mobile Broadband Future of IoT" n.d.). The introduction of 5G will enable a faster and more efficient network for the IoT. Multiple studies have been performed along these lines to enhance the network capabilities of IoT further using 5G(S. Li, Da Xu, and Zhao, n.d.)

4.3 IoT ADOPTION AND APPLICATIONS

Today, IoT devices are available for the automation of day-to-day household-related activities and are also available for large-scale industrial automation. Almost all domains have IoT utilities, including smart household devices, wearable health monitors for healthcare, connected cars in automobiles, traffic sensors for smart cities, and the industrial use of IoT.

GSMA in a survey ("GSMA: The Impact of the Internet of Things" 2019) highlights that the daily usage of IoT devices in the house will increase many-fold. The connectivity costs and the smart device costs continue a downward trajectory, and the benefits attached to them continue to grow; thus, it is very likely that IoT will continue to increase in the coming years. As per the survey, most families owned ten connected devices in 2012. This is likely to increase to 25 in 2017, and up to 50 in 2022 ("GSMA: The Impact of the Internet of Things" 2019).

4.4 IoT FRAMEWORK

There is no well-defined framework of IoT due to a lack of standardization. Hence, multiple models have been proposed (Makhdoom et al. 2019). Some of them are:

- Five-layer architecture – objects or perception layer, object abstraction layer, service management, application layer and business layer (Al-Fuqaha et al. 2015).
- Four-layer architecture – physical layer, perception layer, network layer and the application layer (Kumar, Vealey, and Srivastava 2016).

- Three-layered architecture – sensor, network and application layers (Khari et al. 2016).
- Another five-layered IoT model has been proposed (R. Khan et al. 2012).

However, broadly, all the models define work on three primary layers (Jing et al. 2014) which are:

1. Perception.
2. Transport.
3. Application.

Perception Layer (Frustaci et al. 2018): This layer consists of the sensors deployed and is primarily responsible for data collection. It is the most susceptible layer because of the physical exposure of IoT devices. Moreover, systems in this layer have limitations such as hardware limitations, technological heterogeneity and resource limitations, limiting the implementation of security controls (Amit, Holczer, and Levente 2011). The standard attacks are:

- Node Tampering: Attackers can access the physical node and also tamper with it. Physical nodes and sensors are often placed in vulnerable physical locations. The attackers may replace the entire device with a malicious one or connect to the device to extract information for accessing the data in the layers above.
- Malicious Code Injection: The attacker may tamper with the device's code/application and inject malicious code after gaining physical access.
- Impersonation: Attackers may introduce a new device impersonating a genuine device by exploiting the weak authentication mechanisms.
- Denial of Service: IoT nodes have minimal computational powers. Thus, these can be easily overloaded, making the nodes unavailable.
- Routing Attacks: Intermediate nodes may be exploited to modify the routing paths, leading the collected data to somewhere else or routing erroneous data forward.
- Data Transit Attacks: As the nodes are often deployed in remote locations with a limited network, security is challenging. The system is susceptible to Man in the Middle (MitM) attacks and sniffing.

Transportation Layer (Frustaci et al. 2018): As a result of the weaknesses in the standard wireless communications and threats, this layer, as compared with the perception layer, has a lower risk level. Much research has been done to address the threats at this level, which makes it more secure (Amit et al. 2001).

Common Attacks:

- Routing Attacks: Intermediate nodes may be exploited to modify the routing paths leading the collected data to somewhere else or routing erroneous data forward.
- DoS Attacks: Intermediate nodes may be exploited to modify the routing paths leading the collected data to somewhere else or routing erroneous data forward.

- Data Transit Attacks: As the nodes are often deployed in remote locations with a limited network, security is challenging. The system is vulnerable to MitM attacks and sniffing.

Application Layer (Frustaci et al. 2018): The risk at this level is linked with the application it hosts. Depending on its utility, the availability, integrity and confidentiality can be tolerable or intolerable. This layer has better and mature technology, which makes it much more secure than other layers.

Common Attacks:

- Data Leakage: Vulnerable applications may lead to data loss and loss of confidentiality.
- DoS Attacks: Attackers may attempt to launch an attack causing the application to become unavailable.
- Malicious Code Injection: Vulnerable applications are prone to injection attacks; attackers may inject malicious code.

4.5 IOT SECURITY

IoT security is often neglected for various reasons by manufacturers. Cost-cutting needs and the limited resources in IoT often lead to security being neglected in IoT devices (Frustaci et al. 2018). The deployment of IoT systems in a heterogeneous and often not a secure environment makes them a unique domain with a different security need than the traditional information technology (IT) system.

4.5.1 Lower-level Security Issues

The low-level security problems are related to the physical and data link layer. They include the threat at the physical layer along with the communication at the data link layer (Khan and Salah 2018.) The various security issues are:

- Jamming adversaries.
- Sybil and spoofing (low-level) attacks.
- Insecure physical interface.
- Sleep deprivation attack.

4.5.2 Intermediate-level Security Issues

The intermediate-level security issues are mainly related to network and transport layers, including network communications, routing and sessions management (Khan and Salah 2018) Some of the critical security issues are:

- Replay attacks (from packet fragmentation).
- Insecure neighbour discovery.
- Buffer reservation attack.

- RPL routing attack.
- Sinkhole and wormhole attacks.
- Sybil attacks on intermediate layers.
- Authentication and secure communication.
- End-to-end transport level security.
- Session establishment and resumption.
- Privacy violations.

4.5.3 High-level Security Issues

The high-level security issues are the security concerns related to the applications executing on IoT as described later. The key high level issues are:

Insecure interfaces.
Insecure software/firmware.
Middleware security.

4.6 BLOCKCHAIN

Blockchain contains a sequence of cryptographically interconnected blocks (Makhdoom et al. 2019). Bitcoin is one of the most popular implementations of blockchain. Figure 4.1 shows an example of a blockchain, where each chain consists of multiple transactions. The validation of the blocks is based on cryptographic means. Each block has the hash value of the preceding block, which includes a nonce (a random value). This entire system establishes a trust model wherc each node validates the other nodes and does not require a central authority to validate them.

4.7 BLOCKCHAIN AND IOT

IoT has multiple security challenges that can be overcome using the benefits that the blockchain technology offers. Blockchain provides a unique way to establish trust relationships and immutability of data. IoT's distributed and heterogeneous architecture also stands to benefit from blockchain's decentralized architecture. Hence, blockchain could be one of the solutions to address the security challenges being faced by IoTs today.

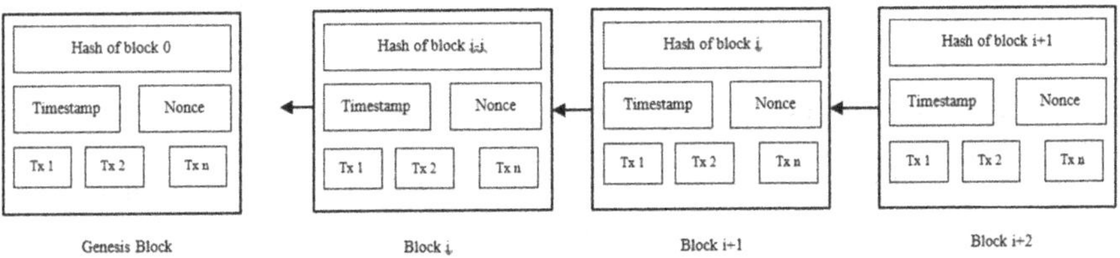

FIGURE 4.1 Blockchain containing a continuous sequence of blocks.

The features of blockchain benefitting IoT implementation are as follows (Khan and Salah 2018):

Address Space: Compared with IPv6 with its 128-bit addressing system, blockchain uses 160 bits for addressing, enabling the scalability of blockchain solutions for IoT, considering the widespread usage of IoT that is likely in the near future (Antonopoulos 2017). IoT devices face challenges in implementing a full IPv6 stack with limited resources and hence, can utilize the blockchain address space.

Governance and Identity of Things (IDoT): Identity and Access Management (IAM) for IoT has several challenges that must be addressed to manage the risks. The device may change hands during its lifetime from the manufacturer to the consumer, including dealers and retailers. The device's ownership needs to be managed seamlessly every time it is sold and resold (Khan and Salah 2018) Blockchain presents an opportunity to address this challenge with its security features and capabilities (Otte et al. 2020).

Data Authentication and Integrity: IoT devices, if connected via a blockchain architecture, would transmit data securely (cryptographically secured) and with trust (digitally signed from sender using its unique public key and GUID or Globally Unique Identifier). Moreover, blockchains provide better auditability of the actions with the ledgers.

Authentication, Authorization and Privacy: Smart contracts can drive a decentralized authentication and authorization in IoTs with a predefined set of rules. Smart contracts are much more effective than the traditional solutions available today.

Secure Communications: MQTT (MQ Telemetry Transport), HTTP (Hypertext Transfer Protocol), XMPP (Extensible Messaging and Presence Protocol) and CoAP (Constrained Application Protocol), including the routing protocols such as 6LoWPAN (IPv6 over Low -Power Wireless Personal Area Networks) and RPL (Routing Protocol for Low-Power and Lossy Networks) used by IoTs, have security concerns by design. These need to be used with protocols such as TLS (Transport Layer Security) or DTLS (Datagram Transport Layer Security) for security enablement. With blockchain, the overhead of ensuring trust can be eliminated; the unique identifiers and the trust model of the blockchain can aid secure communications.

A detailed list of IoT security requirements versus blockchain technology is shown in Table 4.1 (Makhdoom et al. 2019).

4.8 DATA SECURITY IN IOT WITH BLOCKCHAIN

Data security and privacy are among the most sought-out topics today in security. IoT devices, mostly sensors, often deal with sensitive or critical data that could lead to serious security issues if tampered with or lost. Blockchain works on a different trust model, which is decentralized. The full nodes and miners in a blockchain keep

TABLE 4.1
IoT Security Requirements versus Blockchain Technology

Sl. No.	IoT Security Requirements	Blockchain Technology	Sl. No.	IoT Security Requirements	Blockchain Technology
1	Trust free operations	Supported	12	Authorization	Supported using Hyperledger Fabric
2	Distributed architecture	Supported	13	Key management	Supported using Hyperledger Fabric
3	Decentralized deployment	Supported	14	Restricted network access	Supported using Ethereum and Hyperledger Fabric
4	Data Integrity	Supported	15	Device authentication	Supported
5	Data authentication	Supported	16	Software integrity check	Not supported
6	Data confidentiality and privacy	Supported using Hyperledger Fabric	17	Runtime/synchronized software updates	Not supported
7	Pseudonymous IDs	Supported	18	Detection of rogue device	Limited support
8	Privacy	Limited support	19	Consensus protocol for IoT	Not supported

a copy of the blockchain state. Thus, if an adversary could alter the data in any of the nodes, the altered state of blockchain would be automatically rejected, and an alteration would not be possible.

With a distributed architecture, blockchain also ensures the availability of data. Unavailability of some nodes does not disrupt, as replicated data is available in the blockchain environment. Smart contracting can enable a secure way of sharing data with secure authentication and authorization. A predefined set of rules can be defined to ensure that data sharing is secure. "Hyperledger Fabric" is a particular kind of blockchain. It uses a unique execute-order-validate architecture.

4.9 CHALLENGES IN IMPLEMENTING BLOCKCHAIN WITH IoT

Blockchain development was not focused on solving IoT's security challenges, and similarly, IoT has multiple limitations to supporting the blockchain-based implementations. Some of the key challenges are:

Resource limitation: Blockchain is not based on a resource-constrained environment; thus, any solution based on the blockchain must consider the IoT environment's resource constraints. Energy and resource requirement to support multicast and broadcasts, exchange of keys and certificates may not bc suitable for the current IoT architecture. The resource and energy required to support the blockchain operations need to be addressed in IoT.

Heterogeneous devices: The IoT ecosystem with multiple layers has a wide range of systems, including low-power sensors to high-end servers, and is supported by heterogeneous technologies. Blockchain solutions thus need to be technologically conducive to be scalable and popular.

Interoperability of security protocols: IoT has different security requirements at different layers; however, solutions to these must work together to ensure complete security. Thus, a scalable solution can be found if it has wide adaptability.

Trusted updates and management: IoT devices are distributed and can number in the thousands; this poses a severe challenge for updating and trust management. This challenge is one of the open areas for research.

Blockchain vulnerabilities: Blockchains also have known and unknown vulnerabilities (Li et al. 2018). Adversaries may compromise a miner's hashing power. Also, there can only be limited randomness of the private keys.

Physical vulnerabilities: IoT devices are mostly located in vulnerable locations; moreover, they are built with cost-effective, easy to assemble physical parts. This poses an opportunity for adversaries to gain access physically and exploit the underlying layers. Adversaries with physical access may render the security offered by blockchain ineffective.

Lack of IoT-centric consensus protocol: Blockchain consensus protocols are not built specifically for the IoT environment. The absence of a proper consensus protocol may delay the confirmation of the transactions. IoTs are time-sensitive and may not be tolerant of the time delays the current consensus protocols produce.

Transaction Validation Rules: Blockchains have various transaction validation rules; however, most of these were developed to support financial transactions. Generally, the IoT environment consists of heterogeneous devices serving different operations and generating data in different formats with no standardization. These can be challenging to transaction validation rules.

4.10 CURRENT BLOCKCHAIN APPLICATIONS FOR IoT

Scholars and researchers are continuously developing ways to improve blockchains for IoT implementations. Blockchain's advantages of decentralized architecture, trust model, fault tolerance and smart contracts complement IoT's security challenges. Table 4.2 lists some of the blockchain-based applications developed to address the limitations of IoT.

4.11 OPPORTUNITIES AHEAD

There are multiple areas where development is required to address the present-day challenges of blockchain to empower it to address the requirements for IoT integration. Some of the critical areas for future developments are listed here:

IoT-based consensus protocol (Makhdoom et al. 2019): Generally, in the IoT environment, the transactions are not related to the previous transactions. One such case could be reading a sensor, where every reading is independent of the values recorded previously. Thus, for these transactions to be validated, we need a better and contextual to the IoT implementation and environment. Another crucial area of improvement is minimizing the latency of transaction validation.

Fault tolerance: IoT devices are also physically vulnerable, thus providing opportunities for adversaries to introduce or corrupt existing devices into malicious nodes. Thus, a consensus protocol must have the ability to sustain and serve effectively despite malicious nodes in the ecosystem (Makhdoom et al. 2019).

Blockchain size (Makhdoom et al. 2019): Blockchains were not developed to run on resource-constrained systems like IoT. Multiple models have been proposed to address this, such as an off-chain network of private nodes in the form of a Distributed Hash Table (DHT) (Aniello et al. 2017), introducing universal and regional blockchains ("ADEPT: An IoT Practitioner Perspective" 2015), sidechains (Zyskind et al. 2015a, b) and a distributed database of transactions (Gaetani et al. 2017; Aniello et al. 2017).

Scalability: The miner nodes store the complete blockchain and validate the transactions. This storage is one of the critical security features. However, with the increase in transaction volume, this can be a bottleneck, resulting in high latency.

Secure integration of IoT device and blockchain: Smart contracts play an essential role in providing a secure way to integrate authorized IoT devices in the blockchain platform.

TABLE 4.2
Proposed Blockchain Applications for IoT

Application Name	Brief Introduction	Solution Offered	Challenges
ADEPT ("ADEPT: An IoT Practitioner Perspective" 2015)	Provides an autonomous secure framework	It addresses some of the challenges of the IoTs, including addressing the single point failure, data privacy and human intervention errors with an IoT architecture that is secure, scalable, autonomous, robust and decentralized. The proposed solution has peer-to-peer (P2P) messaging using TeleHash protocol and distributed file sharing with BitTorrent.	The current solution is at a proof-of-concept stage and needs to be further strengthened around performance and security concerns.
A security framework for smart cities (Biswas and Muthukkumarasamy 2017)	A blockchain-based security framework for secure communication between smart city entities	The solution provides for a shared platform for secure communication, which can be used by the IoT devices integrated with the blockchain. It provides for data security, including data integrity and availability.	The solutions are yet to be supported by a qualitative and quantitative analysis. The solution is also unclear about the blockchain platform and consensus protocols.
A secure firmware update (Lee and Lee 2017)	Blockchain-based IoT device secure firmware update and integrity check	The solution offers a secure way for devices to check for the latest available firmware, ensure its integrity and download the same. It also addresses the other challenges, including simultaneous download and network load on the bandwidth. It proposes BitTorrent-based peer-to-peer (P2P) file sharing for sharing the firmware.	The solution can be improved to address blockchain forks and transaction confirmation latency. It also needs to address the problems arising from having any security compromise in the network.

(Continued)

TABLE 4.2 (CONTINUED)
Proposed Blockchain Applications for IoT

Application Name	Brief Introduction	Solution Offered	Challenges
VANETS (Leiding et al. 2016)	Decentralized and self-managed VANET (Vehicular Ad hoc Network)	The paper proposes a decentralized and self-managing VANET that is Ethereum blockchain based. VANETS typically are centralized with challenges such as single point failure and data privacy. The solution uses smart contracts to perform the operations guided by rules. The solution also ensures that the nodes self-fund the network infrastructure with Ethers.	The solution does not discuss latency from blockchain and secure communication between vehicles (V2V). Vehicles are expected to make decisions based on near real-time data, and latency in such cases may lead to failure of the system.
Transparency of SCM ("Intel Jumps into Blockchain Technology Storm With 'Sawtooth Lake' Distributed Ledger – Blockchain News, Opinion, TV and Jobs" n.d.)	Object tracking and recordkeeping of ownership	The solution aims to address the challenges of supply chain management (SCM) using blockchain. It maintains a formal registry to track the supply chain from origin to the transformations. The IoT sensors record various aspects of the supply chain, such as locations, storage conditions, temperature, etc., and maintain a ledger.	The solutions do not address trust model issues. The model assumes the reporting nodes to be trusted and provides no means to verify the reported data. However, if the nodes are trusted, then the need for using a blockchain does not arise. Without the implicit trust, none of the nodes can be trusted, and malicious nodes may compromise the entire system.
Enigma (Zyskind et al. 2015b)	Privacy-preserving data computation	The "Enigma" solution aims to address data privacy challenges. The solution provides a secret share of data for the nodes, and no node has access to the other nodes' data. This solution prevents leakage of data among the nodes.	The solution faces challenges from the overheads for communications and computation. The solution also may face challenges from wireless communication regulations.

(Continued)

TABLE 4.2 (CONTINUED)
Proposed Blockchain Applications for IoT

Application Name	Brief Introduction	Solution Offered	Challenges
ELIB (Mohanty et al. 2020)	An efficient lightweight, integrated Blockchain (ELIB) model for IoT security and privacy	The ELIB model proposes an overlay network where IoT devices can securely communicate. It proposes various aspects such as consensus algorithms, DTM (Distributed Throughput Management) and certificate-less cryptography to address the challenges.	The proposed solution can be further expanded to other areas beyond the smart home to examine better the capabilities being offered, and the same can be improved.
IDS (Kim et al. 2018)	Intrusion detection and mitigation system using blockchain analysis for Bitcoin exchange	The solution utilizes the decentralized nature of blockchain for the intrusion detection system (IDS). The proposed solution works with a traditional IDS, and the traditional intrusion prevention system (IPS) is a fail-safe option. This solution addresses the intrusions explicitly in Bitcoin exchanges.	The solution is currently limited to Bitcoin blockchain, and further research needs to be carried out to extend it to other blockchain-based cryptocurrencies. The proposed solution does not do away with traditional IPS and uses it as a fall-back and fail-safe.

Development of blockchain-based IoT communication protocols (Makhdoom et al. 2019): IoT devices are often placed in limited network bandwidth areas. Thus, a protocol needs to be developed, which is customized to support limited bandwidth and is aligned to blockchain's decentralized nature. Multiple peer-to-peer (P2P) protocols have been proposed to enable this ("Telehash – Wikipedia" n.d.).

Secure authentication: IoT devices are physically vulnerable; thus, the chances of malicious nodes are high. Blockchain offers multiple security features, such as smart contracts to enable authentication of valid devices. The blockchains can develop an independent platform that unifies a heterogeneous IoT environment and provides a secure authentication and communication medium (Biswas and Muthukkumarasamy 2017).

4.12 CONCLUSION

The challenges and the opportunities around IoT security have been discussed. With multiple technological and resource limitations in IoT devices, conventional security measures are impractical to implement. However, as discussed, blockchain offers opportunities to address these challenges and also catch up with the technological advancement in this domain. With the research community continuously evolving this subject, blockchain is probably the future of IoT security.

REFERENCES

5G and IoT: The Mobile Broadband Future of IoT. (n.d.) Accessed January 9, 2021. https://www.i-scoop.eu/internet-of-things-guide/5g-iot/.

ADEPT: An IoT Practitioner Perspective. (2015) Accessed January 10, 2021. https://static1.squarespace.com/static/55f73743e4b051cfcc0b02cf/55f73e5ee4b09b2bff5b2eca/55f73e72e4b09b2bff5b3267/1442266738638/IBM-ADEPT-Practictioner-Perspective-Pre-Publication-Draft-7-Jan-2015.pdf%3Fformat%3Dorigina

Al-Fuqaha, Ala, Mohsen Guizani, Mehdi Mohammadi, Mohammed Aledhari, & Moussa Ayyash. (2015). Internet of Things: A Survey on Enabling Technologies, Protocols, and Applications. *IEEE Communications Surveys and Tutorials*, *17*(4), 2347–76. https://doi.org/10.1109/COMST.2015.2444095.

Aniello, Leonardo, Roberto Baldoni, Edoardo Gaetani, Federico Lombardi, Andrea Margheri, & Vladimiro Sassone. (2017). A Prototype Evaluation of a Tamper-Resistant High Performance Blockchain-Based Transaction Log for a Distributed Database. In Proceedings – 2017 13th European Dependable Computing Conference, EDCC 2017 (pp. 151–54). Institute of Electrical and Electronics Engineers Inc. https://doi.org/10.1109/EDCC.2017.31.

Antonopoulos, Andreas M. (2017). Mastering Bitcoin: Programming the Open Blockchain. https://books.google.co.in/books/about/Mastering_Bitcoin.html?id=tponDwAAQBAJ&source=kp_book_description&redir_esc=y.

Atzori, Luigi, Antonio Iera, & Giacomo Morabito. (2010). The Internet of Things: A Survey. *Computer Networks*, *54*(15), 2787–2805. https://doi.org/10.1016/j.comnet.2010.05.010.

Biswas, Kamanashis, & Vallipuram Muthukkumarasamy. (2017). Securing Smart Cities Using Blockchain Technology. In Proceedings – 18th IEEE International Conference on High Performance Computing and Communications, 14th IEEE International Conference on Smart City and 2nd IEEE International Conference on Data Science and Systems, HPCC/SmartCity/DSS 2016 (pp. 1392–93). Institute of Electrical and Electronics Engineers Inc. https://doi.org/10.1109/HPCC-SmartCity-DSS.2016.0198.

Frustaci, Mario, Pasquale Pace, Gianluca Aloi, & Giancarlo Fortino. (2018). Evaluating Critical Security Issues of the IoT World: Present and Future Challenges. *IEEE Internet of Things Journal*, *5*(4), 2483–95. https://doi.org/10.1109/JIOT.2017.2767291.

Gaetani, Edoardo, Leonardo Aniello, Roberto Baldoni, Federico Lombardi, Andrea Margheri, & Vladimiro Sassone. (2017). Blockchain-Based Database to Ensure Data Integrity in Cloud Computing Environments. In Italian Conference on Cybersecurity, Italy.

GSMA: The Impact of the Internet of Things. (2019). https://www.gsma.com/newsroom/wp-content/uploads/15625-Connected-Living-Report.pdf.

Hyperledger Fabric Model: Hyperledger-Fabricdocs Master Documentation. (n.d.). Accessed January 9, 2021. https://hyperledger-fabric.readthedocs.io/en/release-1.2/fabric_model.html#privacy.

Intel Jumps into Blockchain Technology Storm With 'Sawtooth Lake' Distributed Ledger – Blockchain News, Opinion, TV and Jobs. (n.d.). Accessed January 9, 2021. https://www.the-blockchain.com/2016/04/09/intel-jumps-into-blockchain-technology-storm-with-sawtooth-lake-distributed-ledger/.

Jing, Qi, Athanasios V. Vasilakos, Jiafu Wan, Jingwei Lu, & Dechao Qiu. (2014). Security of the Internet of Things: Perspectives and Challenges. *Wireless Networks*, *20*(8), 2481–2501. https://doi.org/10.1007/s11276-014-0761-7.

Khan, Minhaj Ahmad, & Khaled Salah. (2018). IoT Security: Review, Blockchain Solutions, and Open Challenges. *Future Generation Computer Systems*, *82*(May), 395–411. https://doi.org/10.1016/J.FUTURE.2017.11.022.

Khan, Rafiullah, Sarmad Ullah Khan, Rifaqat Zaheer, & Shahid Khan. (2012). "Future Internet: The Internet of Things Architecture, Possible Applications and Key Challenges". In Proceedings – 10th International Conference on Frontiers of Information Technology, FIT 2012 (pp. 257–60). https://doi.org/10.1109/FIT.2012.53.

Khari, Manju, M. Kumar, Sonakshi Vij, Priyank Pandey, & Vaishali. (2016). Internet of Things: Proposed Security Aspects for Digitizing the World. In 3rd International Conference on Computing for Sustainable Global Development (INDIACom).

Kim, Suah, Beomjoong Kim, & Hyoung Joong Kim. (2018). Intrusion Detection and Mitigation System Using Blockchain Analysis for Bitcoin Exchange. In ACM International Conference Proceeding Series (pp. 40–44). New York: Association for Computing Machinery. https://doi.org/10.1145/3291064.3291075.

Kumar, Sathish Alampalayam, Tyler, Vealey, & Harshit, Srivastava. (2016). Security in Internet of Things: Challenges, Solutions and Future Directions. In Proceedings of the Annual Hawaii International Conference on System Sciences (pp. 5772–5781), 2016-March. IEEE Computer Society. https://doi.org/10.1109/HICSS.2016.714.

Lee, Boohyung, & Jong Hyouk Lee. (2017). Blockchain-Based Secure Firmware Update for Embedded Devices in an Internet of Things Environment. *Journal of Supercomputing*, *73*(3), 1152–67. https://doi.org/10.1007/s11227-016-1870-0.

Leiding, Benjamin, Parisa Memarmoshrefi, & Dieter Hogrefe. (2016). Self-Managed and Blockchain-Based Vehicular Ad-Hoc Networks. In UbiComp 2016 Adjunct – Proceedings of the 2016 ACM International Joint Conference on Pervasive and Ubiquitous Computing (pp. 137–40). New York: Association for Computing Machinery, Inc. https://doi.org/10.1145/2968219.2971409.

Li, Shancang, Li Da Xu, & Shanshan Zhao. (n.d.). Internet of Things: A Survey. *Information Systems Frontiers Springer, 17*(2), 243–259, April.

Li, Xiaoqi, Peng Jiang, Ting Chen, Xiapu Luo, & Qiaoyan Wen. (2018). A Survey on the Security of Blockchain Systems. *Future Generation Computer Systems, 107*(February), 841–53. http://arxiv.org/abs/1802.06993.

Makhdoom, Imran, Mehran Abolhasan, Haider Abbas, & Wei Ni. (2019). Blockchain's Adoption in IoT: The Challenges, and a Way Forward. *Journal of Network and Computer Applications, 125*(January), 251–79. https://doi.org/10.1016/j.jnca.2018.10.019.

Mattern, Friedemann, & Christian Floerkemeier. (2010). From the Internet of Computers to the Internet of Things". In *Lecture Notes in Computer Science (Including Subseries Lecture Notes in Artificial Intelligence and Lecture Notes in Bioinformatics)* (Vol. 6462 LNCS, pp. 242–259). Berlin, Heidelberg: Springer. https://doi.org/10.1007/978-3-642-17226-7_15.

Mohanty, Sachi Nandan, K.C. Ramya, S. Sheeba Rani, Deepak Gupta, K. Shankar, S.K. Lakshmanaprabu, & Ashish Khanna. (2020). An Efficient Lightweight Integrated Blockchain (ELIB) Model for IoT Security and Privacy. *Future Generation Computer Systems, 102*(January), 1027–37. https://doi.org/10.1016/j.future.2019.09.050.

Otte, Pim, Martijn de Vos, & Johan Pouwelse. (2020). TrustChain: A Sybil-Resistant Scalable Blockchain. *Future Generation Computer Systems, 107*(June), 770–80. https://doi.org/10.1016/j.future.2017.08.048.

Telehash – Wikipedia. (n.d.). Accessed 9 January 2021. https://en.wikipedia.org/wiki/Telehash.

Uckelmann, Dieter, Mark Harrison, & Florian Michahelles. 2011. An Architectural Approach Towards the Future Internet of Things. In *Architecting the Internet of Things* (pp. 1–24) Berlin Heidelberg: Springer. https://doi.org/10.1007/978-3-642-19157-2_1.

Zyskind, G., Nathan, O., Pentland, A. 2015b. Decentralizing privacy: using blockchain to protect personal data. In: Proceedings of the IEEE Security and Privacy Workshops, pp. 180–184, https://doi.org/10.1109/SPW.2015.27

Zyskind, G., Nathan, O., Pentland, A. 2015b. Enigma: Decentralized computation platform with guaranteed privacy, CoRR abs/1506.03471. http://arxiv.org/abs/ 1506.03471

5 Digitized Land Registration Using Blockchain Technology

P Vinothiyalakshmi, C Muralidharan, Y Mohamed Sirajudeen and R Anitha

CONTENTS

DOI: 10.1201/9781003138082-5

5.1 INTRODUCTION TO BLOCKCHAIN TECHNOLOGY

Blockchain is adistributed ledger technology (DLT) that uses the decentralization method andstores the history of all digital assets.which cannot be altered. With the usage of cryptographic hashing, the blockchain ensures transparent and secure digital transactions. For example, consider the Google Docs concept, where the document can be shared with anyone or to any group but cannot be altered without proper permission from the owner (Benbunan-Fich et.al, 2020). The document can be shared withanyone, and those users can access it with limited access only. Hence, the document is distributed rather than transferred or copied. This distribution is called a*decentralized chain*, which is used in blockchain technology(Ameyaw et.al, 2020). In this model,access to the document can be given to many users at the same time, and no one needs to wait for others to perform the process. But, all the modifications that are made in the document by different users are recorded and transparent to all users. Not exactly likeGoogle Docs, blockchain works in a more complex way, with three critical ideas:

1. Distributed digital assets,which reduces the copying and transferring process.
2. The decentralized access to assets,which allows real-time full access to the asset.
3. Integrity preservation of all the documents through a transparent ledger, thereby creatingtrust over the assets.

With these three categories, blockchain is considered a revolutionary as well as promising technology that reduces risk by avoiding fraudulent activities over the assets. It also ensures transparency of assets with a scalable architecture.

5.1.1 Components of Blockchain

As per the technological review by the Massachusetts Institute of Technology (MIT), the point of people using blockchain technology is that they don't need to trust anyone and the valuable data can be shared in a secure way,i.e., tamperproof.

Blockchain includes six important components:

1. Node –The computer or the user that is present in the blockchain framework. Each node has a separate ledger copy.
2. Transaction –This is the lowestcomponent of the records and other information of theblockchain.
3. Block –This is the component or the data structure where the transaction data of all the nodes that exist in the network is captured. When a block is created,a 32-bit whole number called a nonce will be generated randomly,thereby generating the header of the block.
4. Chain –This is the sequence of all the blocks in a specific order.
5. Miners –The precise nodes that are used toaccomplish the authenticationof blocks before implementing the modification in the blockchain.They

usespecial software for solving complex mathematical problems by finding the suitable nonce that generates the hash.

6. Consensus or consensus protocol – The set of rules for performingblock-chain operations.

5.2 CHARACTERISTICS OF BLOCKCHAIN TECHNOLOGY

The blockchain technology possesses different characteristics, such as the following.

5.2.1 Distributed Ledger

A database that can be shared and synchronized consensually across different websites, geographies and institutions that can be accessed by multiple users is known as a distributed ledger. This database allows all the participants to share the recorded transactions of the networks, and the users can have a copy of the document based on the permissions. All the changes that are made with the document/data will be reflected to the ledger with proper timestamps. This allows the entire system to provide public witness. In contrast to a centralized ledger, the distributed ledger seems to be used by most companies, as the centralized ledger might have the possibility of fraud and cyber-attacksbecause the failover can happen at a single point. Hence, this distributed ledger methodology is used by blockchain technology.

5.2.2 Digitalization

All companies tend to innovate in order to adapt to the dynamic world and to succeed in competitive environments. The current scenario of revolution is towards digitalization, where every organization tends to adapt permanently(Ehmke et.al, 2018). It is mandatory for an organization's business to bypass the traditional models and barriers that exist between the sectors and to change towards the current revolution with the essence of the traditional models. It is also observed that many of the products that were prevalent have now been converted to services without changing the essence of the traditional models. The digital revolution provides the opportunity for the growth of organizations as the consumers tend to move towards the digital environment (Weber et.al, 2019). It also supports the organizations in improving consumer satisfaction by easier observation of the needs of the consumer through direct interaction; this obviously supports the organizations to grow culturally as well as technologically. Blockchain technology handles only digital data or assets over the network.

5.2.3 Updated Near Real-Time

The blockchain is a distributed model where the data/asset that is modified or used by the user will be updated in the respective blocks with proper timestamps. The history of modification as well as the history of the modifier will also be captured and stored in the blocks to make the process more transparent. Thus, the ledger that has been maintained in this distributed technology is more real.

5.2.4 Chronological and Timestamped

The chronological order is the order in which the process of listing, discussing or describing the event is based on its occurrence with respect to time. Blockchain uses timestamping methods, through which a timestamp will be recorded for each and every activity over the document (Yapicioglu et.al, 2020). This makes blockchain work in a secured way, as it allows the user to track all the transactions right from the creation to the modification of the data/document. Since the timestamp is attached to each activity it is possible for the owner to check which event occurred first and what happened later.

5.2.5 Sealed with Cryptography

The blocks that are created in the blockchain will be cryptographically sealed so that it becomes impossible for the user to modify, copy or delete the blocks that are created over the network. This enables a higher level of robustness as well as trust over the data in the blockchain. Further, the decentralized model makes the blockchain failover resilient;a failure happening with a large number of networks might not have an impact on the data, as this model eliminates single point failure. Also, the data stored in the blockchain is immutable (Zheng et.al, 2018).

5.2.6 Irreversible as Well as Auditable

The blockchain has an irreversible property, as a record that is stored in the database over the transaction cannot be reverted back or altered. The records of a single transaction will be linked to every other transaction, which resembles the structure of a chain. It is auditable; as the data is stored with the proper timestamps, each and every process can be audited easily for finding deviations or fraud that occurred over the data.

5.2.7 Transparency

The blockchain is transparent in nature, as every transaction with its associated value can be viewed by the allowed users of the network. Each user or node will have a unique alphanumeric address, which uniquely identifies the user or node. Users have the possibility to remain anonymous or can show their identity to others as a proof. Hence, all transactions will occur between different addresses of the blockchain.

5.2.8 Limited Third Parties

It is acknowledged that blockchain practises fair information,i.e., implementsa set of principles that are related to user concerns and privacy policies. All the transactions can be controlled by the respective user; therefore, the data can be secured through the usage of private and public keys. Since the process is completely secured, it reduces the need forthird-party intermediaries that may misuse the data. Since everything depends on the rules and policies for access, the third party cannot access the data until the owner provides the proper permission.

5.3 PEER-TO-PEER TECHNOLOGY

A peer-to-peer network(P2P) is a decentralized model that includes a group of nodes which save and share recordscollectively; here, each node or device acts as individual peer.The communication in P2P does not have any central control or administration, and all the nodes in the network have equal rights in handling the tasks. This architecture is categorized into three different networks:Structured P2P, unstructured P2P and hybrid P2P networks. In the structured P2P systems, all the nodes or devices will be organized and can search the desired data efficiently in the network. In unstructured P2P systems, the network connection will happen randomly, whereas the hybrid models work by combining both client server and P2P models (Verheye, et.al, 2020). Nowadays, P2P is the foundation for cryptocurrencies and an important part of the blockchain industry.

5.3.1 Peer-to-Peer Network in Blockchain

The P2P technology works with the decentralization model and seems to fit well with blockchain. The P2P architecture of blockchain allows all the cryptocurrencies transferred all around the worldto work with the P2P architecture of blockchains,which works independently without any central or middle man for the control. Those who wish to setup a bitcoin or cryptocurrency can undergo verification and validation of the blocks through the distributed P2P network(Thakur et.al, 2020). In the P2P network, the blockchain seems to be the decentralized ledger for tracing the details of digital assets. Here, if we use the P2P network, all the nodes will be in connection, and each node can have the ledger copy, through which it can compare it with the other nodes for validation of data.

5.3.2 Advantages of P2P in Blockchains

- On comparing with the client server model, the P2P architecture is more secure as there is no central point of control, which obviously avoids complete central point failure. Also,denial-of-serviceoutbreaks are impossible in the P2P network.
- The data has the property of immutability as the data cannot be altered once it is written. Though the network is big, the possibility of alteration is very low, as the majority of nodes will be in connections, which would need to be altered.
- The usage of P2P in a blockchain allows it to run independently without any central point of control.

5.3.3 Limitations of P2P in Blockchains

- The most important and significant disadvantage is that higher computing power is needed for this type of architecture because the nodes are distributed and have no central control.
- Lack of widespread adoption and scalability.

5.4 USAGE OF BLOCKCHAIN TECHNOLOGY IN LAND REGISTRATION

One of the most arduous aspects of the real estate sector is the registration of land, which seems to be crucial for most people except for those who carry out the process themselves(Banupriya S and Arunkumar S, 2020).It is one of the mundane administrative matters,involving amanual stamping process,when every person is excited about getting the key for their new home. Also, the registration process cannot be understood by many people. During 2018, it is calculated that real estate transactionsexceeded $1.7 trillion,4%higher than in 2017.

Also, there will be a land registration trial paper,which will vary both quantitatively and qualitatively.

5.4.1 Challenges of the Land Registration Process

Usually, the registries that are kept for entering the land details will record the ownership details in physical form, as happened in past decades.We might think that it is a simple task, but it includes myriad challenges:Ifsafeguarding is too hard,the record may be lost or destroyed, falsified or manipulated (Kaczorowska et.al, 2019). In the UK, the land registration process has become digital; however, if someone needs to sell a property and doesn't have the document,he needs to prove that he is the owner through proper claims. This process can result in an unfortunate delay. Also, if a disaster destroys the place where the physical documents are maintained, the complete process will collapse. In 2010, a serious earthquake affected Haiti,resulting in more than 1 million people losing their lives (Kamkar et.al, 2019). In such cases, they cannot prove their ownership of property. In these situations, we need to maintain a distributed digital ledger, which might help in solving these problems.

5.4.2 Blockchain on Land Registration

Blockchain technologyis a possible and effective solution for overcoming the challenges in the land registration process. The implementation of land registration using blockchain will record the details of the land as well as the owner and will be assigned to the particular owner's account Krishnapriya et.al, 2020). Any update on the particular land can be updated in a timely way; for example, if a building is built on the empty land, this can be updated in the particular owner's account, or if the property is sold,it can be transferred to the respective account, and so on. Thus, each and every transaction can be traced with a proper timestamp and is indisputable (Kshetri et.al, 2018). This enables the blockchain technology to provide more safetyfor records of ownership. Since the documents are not stored in a central registry, disaster or any other failure will not affect the system, as the data will be dispersed in a distributed manner. With the fingerprint of the particular user, we can replicate it and avoid fake entries over the particular land.

Thus, the paper-based system will not be flexible, durable or resilient.

5.5 EXISTING LAND REGISTRATION SYSTEM

Land registration is a system whereby property rights are recorded and evidence is maintainedin relation to property title holders. In the existing land registration system:

1. The seller and the buyer are included in the land trade negotiation in the sale.
2. If the buyer gets a loan from a bank, the seller needs to provide a certificate stating that the land is not involved in any legal or monetary liabilities.
3. The seller and the buyer get the legal opinion on the prepared sale deed.
4. After the sale deed verification, the stamp duty is paid based on the government's guidelines to the treasury, and stamp paper is bought.
5. The sale deed is signed by both the seller and the buyer when the agreed amount is paid to the seller by the buyer.
6. After validation and verification, the sale deed is registered at the Sub Registraroffice.

Finding the current property holder is challenging among thousands of land records. Some of the issues in the existing land registry system are:

1. The existing land records are not well maintained and are ambiguous. Compared with reality, there are discrepancies in the recorded data.
2. Before sealing the transaction, the submitted documents need to be verified. For the verification process, a team is required to verify the authenticity of the submitteddocuments.
3. The existing land registry system has time-consuming processes.
4. Chance of human error.
5. Failure to detect forgery.
6. The system is not available to keep track of the credit on the land.
7. A third party's involvement.
8. There may be a chance for hiding and manipulating details during registration.

Major challenges in the existing land registry system are:

1. Intrusion by brokers and middlemen

   Brokers and middlemen are an integral part of the land registration system who know more about market offerings. Sellers and buyers prefer to call the brokers and middlemen for their side process completion. Middlemen or brokers gather the required information from traders, lenders, intermediaries, etc., which leads to additional costs, making the entire land registry system an expensive system(Singh et.al, 2020).
2. Number of fraud cases

There may be a chance of imposters posing as the seller of a property. In many situations, both the buyers and the sellers have been unaware of the fraud until it was discovered by the land registry as part of a spot check exercise.

3. Human error

   The land registry system is more vulnerable to human error. Because the updates to the land registry records are done manually, and the accuracy of these updates depends on a particular individual, human intervention leads to the chance of errors in the land registry system.

4. Time delays

   The land registry takes several months for registration and updating. Many legal issues delay the entire process,and buyers have to wait for a long time.

5.6 DIGITAL LAND REGISTRATION SYSTEM

The digitalized land registration system provides confidentiality and security and reduces the time and the labour involvement in selling property. To develop a land registration system digitally, re-engineering procedures are required to permit electronic submission of records and authenticate the legitimacy of those records. Transformationis needed mutually atthe technical level and the legal level (Shang et.al, 2019).The digitalized land registration system improves the turnaround time for carrying out business at the Deeds Registry(Eder G,2019). This ensures that land information is accessible and transparent to the public. The overall architecture of the digital land registration process is shown in Figure 5.1. Advantages of a digitalized land registration system are:

1. It is possible to unitea huge amount of information in a single database.
2. It helps to minimizereplication in the information storage.
3. It optimizes the methods by reorganization workflows and helps in compiling information in ways that arenot possible withmanual systems.
4. It speeds up the processing time in transferring property rights.
5. It is used to set up tracking mechanisms in a land registrysystem to assess its performance and improvethe services to the customers.
6. It provides data accuracy.
7. It keeps the informationup to date.
8. It provides a built-in method for quality control.
9. Itenables consistency checks and data verification to be achievedpromptly.
10. It increases security by allowing backup copies to be made available. The latest data can be saved in different locations, and data are protected from natural disasters.
11. It provides data transparency by creating land records available to the public.
12. People can access data in diverse locations simultaneously.

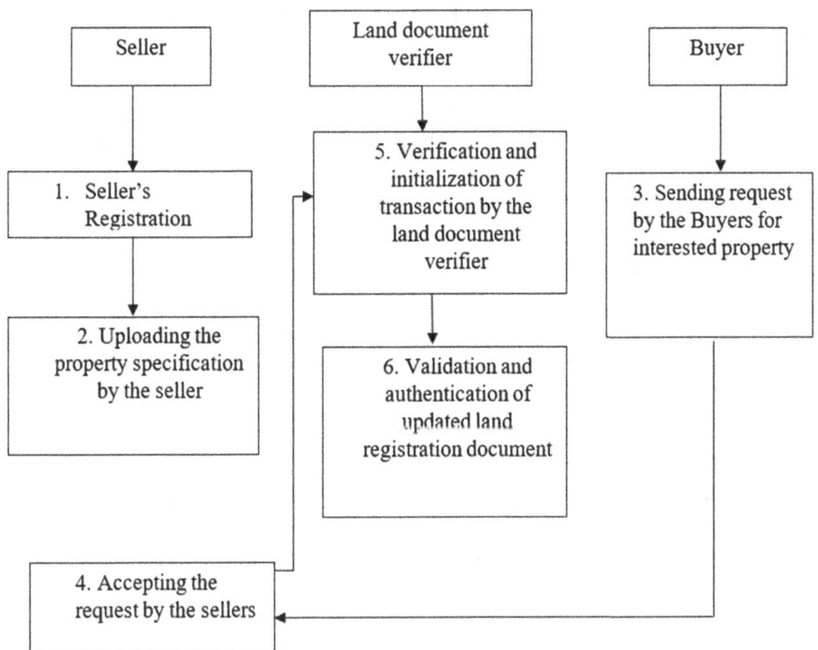

FIGURE 5.1 Architecture of digitalized blockchain technology-enabled land registration system.

Selecting suitable technology is an importantphase in developing a new digital system. Various phases of development needvaried technological solutions that take care of any restraints and limits. In India, land-associated fraud is a major issue. Plenty of cases of forgery and fraud related to land credit are reported. The blockchain technology is a structure that comprisestransactional records and offers transparency and security in a distributed environment (Alam et.al, 2020). It exists in a distributed fashion, where the transactional records are considered to be a chain in which no user acquires authority over the blockchain. The transactional records in blockchainare accessibleto all the participants in the network (Norta et.al, 2018). The originality of the transaction is proved by verifying the digital signature, which is in an encrypted format.

The challengesthat are overcome by blockchain technology in the existing land registry system are:

1. Accelerating the process

   The blockchain land registry framework offers a disseminated database in whichanybody can record and access information without the participation of any central authority. Creating a digital title with a blockchain land registry platform can improve the accuracy of the recording details (Pawar et.al, 2020). Due to the blockchain's power to demonstrate authenticity, proprietors can transmit the land ownership legally to the purchaser without requiring third-party authentication.

2. Reducing fraud cases

   It is possible for imposters to pose as the vendor of a property with expurgation software. In a blockchain land registry system, the seller can upload the sale deed document with their digital signature in the blockchain network, and this can be verified by the user whenever needed (Saranya et.a l, 2019). Because of its incontrovertible property, the blockchain proves the land owner's title and prevents fake documents.
3. Transparency with smart contracts

   The land registry takes several months for registration.By automating verified transactions, smart contracts enablea simple and efficient process. The ownership transfer processesare quicker than the process of the traditional system.Smart contracts makeupdating ownership simpler and quicker, and the changes are stored on the blockchain.
4. Accuracy
5. Updates to the land registry records are automated, and there is less human intervention, which reduces the probabilities of mistakes in the land registry system.

Therefore, the blockchain platform for land registry mightassist asproof of ownership, presence, interchange and transactions.Blockchain maintains all kinds of record keeping more efficientlywith the huge amount of publicly accessiblerecords.

In a digitalized land registration system using blockchain technology,

(i) Data are distributed, and these distributed nodes are connected via a net.
(ii) The record of transactions is transparent to the public.
(iii) The record of transactions is immutable in nature (not changeable).

Blockchain technology also has many features that lead to integrity, reliability and efficiency in a digitalized land registration system.The architecture of a digitalized blockchain technology-enabled land registration system is shown in Figure 5.1.

5.6.1 Actors of Digitalized Blockchain Technology-enabled Land Registration System

The actors of a digitalizedland registration system using blockchain technology are:

1. Property holder or seller – A person who currently holds the property and is ready to sell the property to buyers.
2. Buyer – A person who is ready to interact with the property holder to buy their property.
3. Land document verifier– A person who has the authority to verify the land documents and initiate the transaction.

5.7 WORKING PRINCIPLES OF DIGITALIZED LAND REGISTRATION SYSTEM USING BLOCKCHAIN TECHNOLOGY

(i) Users' registration

People who are ready to buy or sell the property first register their information in the blockchain environment. In this phase, the user needs to register themselves with their details in the registry server, and the registry server will verify the user's details for authentication. If the verification is successful, the registry server generates the user-ID for the particular user and sends this information to the user for login to their account. Then, the user's information is hashed and stored in the blockchain environment.

(ii) Uploading the property specification by the seller

The seller can save their property's documents, images and land locations on theblockchain. In this phase, the seller generates the digital form, which includes the land details such as area coverage, price of the land, etc. The seller has to upload the sale deed with their digital sign in the blockchain environment. If the property details are saved successfully, then all the buyers can view them.

(iii) Sending request by the buyers interested in the property

If the buyer is interested in the property, he verifies the sale deed with a digital sign. Then,he sends the request to the corresponding property holder, and the property holder may accept or reject the request by viewing the buyer's details.

(iv) Accepting the request by the seller

If the seller accepts the property transaction request, the notification will be sent to the land document verifier.

(v) Verification and initialization of transaction by the land document verifier

If the land document verifier receives the notification, the land document verifier verifies the land documents and initiates the transaction from seller to buyer by scheduling a meeting with them.

(vi) Validation and authentication of updated land registration document.

The land document verifier validates the documents submitted by both seller and buyer and then adds the validated records on the blockchain land registry system. Then, infront of the land document verifier, the seller and buyer sign the property ownership transfer document, which will be uploaded on the blockchain land registry system.

Finally, the buyer will pay the amount to the seller for property transfer. .Then, the document is authenticated and modification is not possible in that document.

5.8 CONCLUSION

The land registration system should protect and keep updated propertyinformation to ensure title certainty. A centralized and paper-based system can lead to the loss of rights due to a natural disaster. The digitizing land registration system also has some risks, such as the alteration of records, error or vulnerability to fraud. These kinds of riskshave affected the integrity of stored records.The blockchain technology improves and digitalizes the central registry process to build the underlying trust infrastructure. It improves transparency and prevents fraud cases. It improves-the data security, reliability and legitimacy of land records. Blockchain technology has many characteristics, such as immutability, decentralization, transparency and timestamp features, etc., that provide further certainty over the stored contents of the land registry system.A complete record is mandatory to avoid doubts and uncertainties. Thus, blockchain technology will provide a reliable digitalized land registration system with certainty and a true picture of the rights that are held by a person over a specific property.Future work could be the implementation of a hybrid blockchain and the outline of land-related funding, security simulation and prevention. Other important future work could include smart contract security, risk analysis and Public Key Infrastructure management for secure electronic transfer of information.

REFERENCES

Alam, K.M., Ashfiqur Rahman, J.M., Tasnim, A., & Akther, A. (2020). A Blockchain-based land title management system for Bangladesh. *Journal of King Saud University: Computer and Information Sciences*. 1–15. DOI: 10.1016/j.jksuci.2020.10.011

Ameyaw, P.D., &de Vries, W.T. (2020). Transparency of land administration and the role of blockchain technology, a four-dimensional framework analysis from the Ghanaian land perspective. *Land*, *9*(12), 491.

Banupriya, S, &Arunkumar, S. (2020). Land registration using smart lending contract (slc) in blockchain. *International Journal of Advanced Science and Technology*, *29*(6), 4787–4795.

Benbunan-Fich, R., &Castellanosm, A. (2018). *Digitalization of land records: From paper to blockchain*. ResearchGate.

EderG. (2019). Digital transformation: Blockchain and land titles. In Proceedings of the OECD Global Anti-Corruption & Integrity Forum (pp. 20–21), Paris, France.

Ehmke, C., Wessling, F., &Friedrich, C.M. (2018). Proof-of-property: A lightweight and scalable blockchain protocol. In IEEE/ACM 1st International Workshop on Emerging Trends in Software Engineering for Blockchain (WETSEB). https://ieeexplore.ieee.org/document/8445059/

Kaczorowska, M. (2019). Blockchain-based land registration: Possibilities and challenges. *Masaryk University Journal of Law and Technology ,13(2):339*. DOI: 10.5817/MUJLT2019-2-8

Kamkar, R., Adak, S., Khan, A., Pandit, P., &Sampath, A.K. (2019). Land registry using block chain. *International Journal for Scientific Research & Development*, *6*,(11), 636–639.

Khan, R., Ansari, S., Sachdeva, S., &Jain, S. (2020). Blockchain based land registry system using Ethereum Blockchain. *Journal of Xi'an University of Architecture & Technology*, *12*, 3640–3648.

KrishnapriyaS, &Greeshma, S. (2020). Securing land registration using blockchain. *Procedia Computer Science*, *171*, 1708–1715.

Kshetri, N., &Voas, J. (2018). Blockchain in developing countries. *IEEE IT Professional*, *20*(2), 27–30.

Mertz, L. (2018). A blockchain revolution sweeps into health care, offering the possibility for a much-needed data solution. *IEEE Pulse*, *9*(3), 4–7. DOI: 10.1109/MPUL.2018.2814879

Norta, Alex, Chad Fernandez, & Stefan Hickmott. (2018) On blockchain application: Hyperledger fabric and ethereum. Commercial property tokenizing with smart contracts. In 2018 International Joint Conference on Neural Networks (IJCNN). IEEE, Brazil.

Pawar, R.R., & Chillarge, G.R. (2020). property registration and ownership transfer using blockchain. *International Journal of Innovative Technology and Exploring Engineering (IJITEE)*, *9*(6), 900–903.

Saranya, A., & Mythili, R. (2019). A survey on blockchain based smart applications. *International Journal of Science and Research (IJSR)*, *8*, 450–455.

Shang, Q., & Price, A. (2019). A blockchain-based land titling project in the republic of Georgia: Rebuilding public trust and lessons for future pilot projects. *Innovations Technology Governance Globalization*, *4*, 72–78.

Singh P. (2020). Role of blockchain technology in digitization of land records in Indian scenario. *IOP Conference Series: Earth and Environmental Science*, *614*, 1–13. DOI. 10.1088/1755-1315/614/1/012055

Thakur, V., Doja, M.N., Dwivedi, Y.K., Ahmad, T., &Khadanga, G. (2020). Land records on Blockchain for implementation of land titling in India. *International Journal of Information Management*, *52*, 101940.

Verheye, B. (2020). Land Registration in the Twenty-First Century: Blockchain Land Registers from a Civil Law Perspective. In Lehavi A., Levine-Schnur R. (Eds.), *Disruptive Technology, Legal Innovation, and the Future of Real Estate*. Springer, Cham. https://doi.org/10.1007/978-3-030-52387-9_7

Weber, I., Lu, Q., Tran, A.B., Deshmukh, A., Gorski, M., &Strazds, M. (2019). A platform architecture for multi-tenant Blockchain-based system. IEEE International Conference on Software Architecture (ICSA2019). https://arxiv.org/abs/1901.11219. DOI: 10.1109/ICSA.2019.00019

Yapicioglu, B., & Leshinsky, R. (2020). Blockchain as a tool for land rights: Ownership of land in Cyprus. *Journal of Property, Planning and Environmental Law*, *12*(2), 171–182. DOI: 10.1108/jppel-02-2020-0010.

Zheng, Z., Xie, S., Dai, H.N., Chen, X., & Wang, H. (2018). Blockchain challenges and opportunities: A survey. *International Journal Web and Grid Services*, *14*(4), 352–375.

6 Towards a Regulatory Framework for the Adoption and Use of Cryptocurrencies in Zimbabwe

Tinashe Mazorodze, Vusumuzi Sibanda and Ruramayi Tadu

CONTENTS

DOI: 10.1201/9781003138082-6

6.1 INTRODUCTION AND BACKGROUND

Bitcoin and cryptocurrencies have over the past decade experienced a rapid rise in value and as such, attracted investors from all sectors of society. Since its introduction slightly over a decade ago, Bitcoin has encountered a phenomenal rate of growth in value, peaking at 1 bitcoin being the equivalent of almost 20,000 USD and in the process inspiring thousands of other cryptocurrencies. The year 2013 experienced the start of accelerated growth in cryptocurrencies from a modest 66 varieties to 644 in 2016, reaching 1,335 at the end of 2017, and further hitting 2,116 in January 2019 (Rochemont and Ward, 2019). A similar pattern is evident in market capitalization, where crypto-assets have grown exponentially from around USD 10 billion at the end of 2013 to USD 572.9 billion at the end of 2017 (Center for the Governance of Change [CGC], 2019). This has spurred a significant proportion of the members of the public and institutional investors to invest and trade in cryptocurrencies. The International Data Corporation (IDC) further forecasts that by 2022, universal expenditure on blockchain technologies will exceed $11 billion (Goepfert, 2018).

Zimbabwe, as part of the global village, has not been spared this frenzy. As with all emerging technologies, they come with risk factors coupled with beneficial ones.

> Money is a tool that has been shaped with the development of society, that is, a greater ability to grow and adapt to the nature of the times. Not surprisingly, the money came out of the latest technological developments and, in particular, the widespread use of the Internet.
>
> **(European Central Bank, 2015)**

Technology has thus resulted in the evolution of money, initially by the use of electronic representation of fiat money, such as credit cards and debit cards. The subject of this study on cryptocurrencies, however, goes a step further, in that cryptocurrencies do not represent a paper-based currency but have value in themselves based on trust in the blockchain ledger system. Adoption of cryptocurrencies by Zimbabweans remains at low levels, yet there are several benefits of using cryptocurrencies that may accrue to both the individual users and the country as a whole. Furthermore, there are many doubts as to the legitimacy of cryptocurrencies, especially in the African context; hence the desire through this study to explore issues of cryptocurrency regulation to legitimize it and improve its uptake and acceptance.

While Bitcoins and cryptocurrencies have been widely embraced by businesses, traders and investors, among others, they have, however, been treated with much scepticism by many governments and regulatory authorities. The hazy regulatory environment cannot stop the acceptance frenzy that has been triggered by Bitcoin/cryptocurrencies. There is indeed a growing and irresistible interest in embracing this new technology, with most businesses taking the lead (Yazbeck, 2007; CGC, 2019). Zimbabwe's economy is quite unique, with its monetary system having collapsed twice in a space of two decades. In this vein, cryptocurrencies present a compelling option that may meaningfully address the current cash crisis, which has been a strong catalyst in the use of electronic money such as Point of Sale machines and mobile money technologies. Against this background, it is important to question and interrogate whether cryptocurrencies can be a relevant option given an enabling regulatory environment. Pursuant to answering these questions, the main purpose of this chapter is to proffer regulatory solutions that would encourage and promote the adoption and use of cryptocurrencies in Zimbabwe. This was achieved by investigating the risk factors of cryptocurrency usage and proposing mitigatory measures. The study adopted a mixed methods approach to investigate this phenomenon. Quantitative data collection tools were primarily used to collect data on the adoption and utilization of cryptocurrency currencies, while qualitative data was extracted from key informant interviewees (KII), focusing more on interrogating best practices in regulating cryptocurrencies.

6.2 LITERATURE REVIEW

6.2.1 An Overview of Cryptocurrency

A cryptocurrency, also known as a "digital currency" or "virtual currency", is a symbol in a distributed consensus ledger (DCL) that signifies a component of an account. A cryptocurrency is gained, deposited, utilized and executed by electronic means. It enables peer-to-peer exchange without a third-party intermediary (Accenture Consulting, 2017). Cryptocurrency is not exchangeable with any other service, like gold, though cryptocurrency has no physical form and is not supported by any legal organization. Furthermore, its supply cannot be analyzed by any bank, and it has a decentralized network system, where every transaction is carried out by the users (PwC August, 2015). For a digital currency such as cryptocurrency to operate, it requires a platform such as blockchain (PwC March, 2018). A blockchain is a distributed ledger technology (DLT) that permits data to be kept on servers universally while allowing anybody on the network to perceive everybody else's real-time entries, which enables participants to confirm transactions without the need for a central certifying authority (PwC May, 2017).

6.2.2 Cryptocurrencies and Capitalization

The inception of Bitcoin in 2009 has since triggered the emergence of over 1,600 other cryptocurrencies (Hileman and Rauchs, 2017). According to coinmarketcap.com, a website dedicated to monitoring the Bitcoin ecosystem, as of 1 May 2018, 1,593 different cryptocurrencies were in existence with a total market capitalization of USD 382,816,049,760. The majority of these cryptocurrencies are anchored by the blockchain mechanism, though they naturally exist on remote transaction networks. Numerous copies of Bitcoin have been developed from more important inventions of the empowering blockchain technology, although with diverse constraints like materials, transaction authentication times, etc. However, in terms of market capitalization, the cryptocurrency ecosystem is still dominated by a few players, with Bitcoin accounting for more than 35% of the total market by capitalization. In terms of trading platforms for crypto-assets, as of April 2018, the number had exceeded 10,000 (Rochemont and Ward, 2019). The ten largest cryptocurrencies account for 77.4%, totalling $296,596,186,406 in value, of the total market capitalization, as illustrated in Figure 6.1.

Mourdoukoutas (2018) lists the most popular cryptocurrencies by market capitalization as Bitcoin, Ethereum, Ripple and Litecoin. Based on previous surveys, 76% of millennials would invest in Bitcoin, with the remainder split in half between Ethereum and Litecoin. Bajpai (2017) lists six other altcoins, which could in the future stand as strong alternatives to Bitcoin. These are Litecoin, Ethereum, Dash, Ripple, Monero and ZCash.

6.2.3 Barriers and Opportunities in Cryptocurrencies

Despite its phenomenal growth, cryptocurrency remains saddled with potential threats of money laundering schemes, cyber theft, tax evasion, bribery payments and funding of counterfeit goods as its major dark side (Regulatory Brief-PwC, 2018). Furthermore, cryptocurrencies are deemed potentially disruptive to the financial

▲ #	Name	Symbol	Market Cap	Price
1	Bitcoin	BTC	$142,129,014,515	$8,338.77
2	Ethereum	ETH	$70,656,327,685	$709.73
3	Ripple	XRP	$26,845,167,434	$0.685001
4	Bitcoin Cash	BCH	$21,032,359,782	$1,227.23
5	EOS	EOS	$11,823,809,554	$13.61
6	Litecoin	LTC	$7,739,477,733	$136.70
7	Cardano	ADA	$6,376,140,749	$0.245926
8	Stellar	XLM	$6,009,999,525	$0.323517
9	IOTA	MIOTA	$4,912,013,711	$1.77
10	TRON	TRX	$4,819,875,718	$0.073308

FIGURE 6.1 Cryptocurrencies by market capitalization. (From coinmarketcap.com, 2018)

markets as they are outside the purview of typical central banks and related financial sector regulators (Accenture Consulting, 2017).

However, there is quite a handful of cryptocurrency benefits, which include possible economic effectiveness; being a substitute for prevailing mediators and organizations; acting as an enabler of mobile and digital commerce; ensuring stability in the financial system; working as a crypto-reserve currency; effectively monitoring the supply of money; lowering transaction costs, especially for cross-border transactions; and also allowing traceability (Accenture Consulting, 2017; Schrodt, 2018). Those who support cryptocurrencies preach that a decentralized payment system working over the internet will be less expensive than the old-style payment systems and present organizations (Schrodt, 2018). For those who lack confidence in the adequacy of the traditional financial systems, cryptocurrencies could serve as a worthwhile alternative (Bershidsky, 2017). In developed countries, the level of scepticism may not be as pronounced as in less developed economies with emerging financial and capital markets. Classically, advanced economies are comparatively constant and have moderately low inflation; frequently, they also have carefully controlled financial organizations and robust government organizations. Therefore, cryptocurrencies experience more extensive acceptance in countries with an advanced degree of distrust in present systems than in countries where there is usually a high degree of faith in current systems (Chun, 2017). However, some critics argue that the benefits of cryptocurrency are obscure, as they are more futuristic than immediate.

6.2.4 The Regulation of Cryptocurrencies

The regulation of cryptocurrencies still remains an issue of concern in the African continent and the world in general. Different and cautious approaches have been adopted, with China, for instance, completely banning cryptocurrencies as of 2017, while Switzerland has embraced them with an abundance of caution. Other leading economies, such as Japan and the UK, are reportedly half-hearted. The common thrust amongst all regulators has been their concern to monitor the markets to prevent theft, fraud, market manipulation and money laundering (Regulatory Brief-PwC, 2018). Equally, African countries have adopted blockchain and cryptocurrencies in mixed proportions, with most governments being apprehensive, reserved and applying an abundance of caution in their approaches. Nations like Zimbabwe, Zambia, Swaziland and Namibia have commenced with a cautious stance, with Mauritius being the local pacesetter (McKenzie, 2018). While there is no official position on cryptocurrencies in Botswana, there are, however, traces of its existence and operation in certain circles largely limited to WhatsApp and Facebook groups (McKenzie, 2018). In countries such as Ghana, Kenya and Nigeria, cryptocurrencies are neither recognized nor supported. In particular, the central bank of Nigeria is reported to have cautioned financial institutions in 2017 against using, holding or trading virtual currencies pending substantive regulation or decision; they are not legal in Nigeria. Furthermore, the South African Reserve Bank also does not recognize cryptocurrency as legal tender or currency. The South African Revenue Service (SARS) has also echoed the same sentiments by stating that cryptocurrencies are not authorized

South African tender and are also not extensively utilized and known as a medium of exchange. Consequently, SARS stands ready to punish those who evade tax through cryptocurrency assets.

Despite all this, cryptocurrency continues to experience phenomenal growth and acceptance. It has been embraced by a serious number of stockholders, engineers, controllers, dealers, tycoons and customers (US Congressional Research Service, 2015). Consequently, most countries have been caught off-guard and are now playing catch-up in a bid to protect the investing public and maintain market stability while being cautious enough not to stifle innovation (Regulatory Brief-PwC, 2018). In April 2020, South Africa made a U-turn in promulgating cryptocurrency laws by issuing an outline application, and currently, Nigeria plans to control cryptocurrencies among its Securities and Exchange Commission (Helms, 2020; Kazeem, 2020). Consequently, most attention worldwide is shifting towards the regulation of cryptocurrencies rather than whether to accept them or not.

6.3 METHODOLOGY

This work has chosen a mixed method technique combining both qualitative and quantitative research methods as proposed by Hulme (2007), Greener (2008) and Creswell (2014). This study sought to analyse how best Zimbabwe can regulate the adoption and use of cryptocurrencies. The quantitative approach adopted methods such as surveys and questionnaires with a set of questions and predefined answers (Saunders et al., 2016). The two methods complemented each other in the analysis of data.

6.3.1 Target Population

According to cryptocurrencies network exchange players, the Zimbabwe cryptocurrency database has 10,000 cryptocurrency users either transacting or constantly in touch with the Bitcoin Exchanges, and these cryptocurrency users formed the core of the study population. The population thus consisted of 10,000 cryptocurrency users, two Bitcoin exchange managers, one official from the Banker's Association of Zimbabwe (BAZ), one official from the Reserve Bank of Zimbabwe (RBZ), one Bitcoin entrepreneur and one academic/technologist. Table 6.1 indicates the population and sample composition used for data collection.

6.3.2 Sampling Technique

To select cryptocurrency users, judgmental sampling was used (Kumar, 2009). The researcher relied on his experience as a player in the technology arena to select cryptocurrency users who had a record of having an interest in constantly trading in cryptocurrencies in the Zimbabwean market. The researcher deliberately targeted respondents who were likely to supply this particular study with the required information (Singleton and Straits, 2018; Leedy and Ormond, 2014). For the qualitative aspect of the study, interviewees were selected through convenience sampling (Castillo, 2005).

TABLE 6.1
Population and Sample Composition

Category	Target population	Sample size	Research tools
Cryptocurrency users	10,000	384	Questionnaires
Bitcoin exchange managers	2	2	Meetings
BAZ official	1	1	Meetings
RBZ official	1	1	Meetings
Bitcoin entrepreneur	1	1	Meetings
Academic/technologist	1	1	Meetings
Total	**10,006**	390	

Source: Primary data.

6.3.3 Sample Size

The sample size is chosen using the following formula:

$$\text{Sample Size} = \frac{\dfrac{z^2 \times p(1-p)}{e^2}}{1 + \dfrac{z^2 \times p(1-p)}{e^2 N}} \tag{6.1}$$

The sample size for cryptocurrency users in this study was estimated by adapting a simplified formula for calculating sample size from Yount (2006), where N is the population size, n is the sample size, and $e = 0.05$ level of precision at 95% confidence level. By applying this formula, the sample size was estimated as $n = 384$.

However, to determine the sample size for Bitcoin exchange managers, the BAZ official, the RBZ official and academics in Zimbabwe, the scholar applied the 100% basic rule as recommended by Yount (2006), who proposes a 100% lowest sample size for populations smaller than 10 units. Therefore, the sample size contained two Bitcoin exchange managers, one BAZ official, one RBZ official and two academics/entrepreneurs, as depicted in Table 6.1.

6.3.4 Research Instruments

For the quantitative aspect, questionnaires were administered online (Cooper and Schindler, 2014), coupled with interviews with Blockchain exchange managers, BAZ officials, RBZ users and academics to cover the qualitative aspect (Sekeran, 2009). To ensure validity, the researcher ensured that the survey addressed the themes emphasized in the aims, and examined the data and finished references on the basis of these themes (Gray, 2006). On the other hand, to ensure reliability, the researcher adhered to recommended sample sizes as well as ensuring that the research was completed within a reasonable period of time (Saunders et al., 2016). Ethical features mentioned by the researchers comprised searching and receiving advice on conducting the research and consideration for persons' liberty, volunteerism and privacy, as suggested by Christensen et al. (2011).

6.4 DISCOVERIES AND DELIBERATIONS

6.4.1 Response Rate and Demographic Information

Of the 384 questionnaires administered, 140 respondents completed the questionnaire, bringing the response rate to 36.5%. On the face of it, this appears to be a low response rate; however, according to Saunders et al. (2016), a 35% response rate is satisfactory for most academic studies, with other prior online surveys recording a response rate as low as 20% (Alrousan & Jones, 2016). The biographic information of respondents included age, academic qualifications, gender and employment status as well as level of income. The majority of the respondents were aged between 26 and 35 years (57.2%), followed by the age group of 36-45 years (23.6%); thus, 80.8% were in the able-bodied working group. The older-aged group had a smaller number of respondents, represented by a combined 5.7%. Based on this outcome, it may be argued that most of the respondents were youths, and the study targeted this group. With respect to educational qualifications, the majority of the respondents (40.8%) had a degree, followed by diploma holders at 27.1%, while the postgraduate group accounted for 24.3%. The number of respondents with secondary-level education was insignificant, accounting for a mere 7.9%. This has been studied by several researchers, such as Tornatzky and Fleisher (1990), whose findings show that education does influence the validity of the findings. Lastly, regarding employment status, most respondents (63.6%) were in full-time employment, in contrast to the 90% unemployment rate in the country (ZimStats, 2018).

6.4.2 Most Preferred Investment

The last issue tackled under demographics was the form of investment respondents had in mind. Table 6.2 indicates the distribution of some of the investment options

TABLE 6.2
Other Forms of Investment

	Frequency	Percentage	Valid Percentage	Cumulative Percentage
Stock exchange	31	22.1	22.1	22.1
Unit trusts	9	6.4	6.4	28.5
Savings	66	47.1	47.1	75.6
Properties	9	6.4	6.4	82
Business	22	15.8	15.8	97.8
Other	3	2.2	2.2	100
Total	**140**	**100.0**	**100.0**	

Source: Primary data.

under consideration, which included stock exchange, unit trusts, savings, properties and business, amongst other options.

The main investment portfolio in the harsh Zimbabwean economic environment was savings, as represented by 47.1% of the respondents. Trading on the stock market was mentioned by a huge 22.1% of respondents, enough to suggest that there would be an interest in investment. The other discussed portfolio was business. People try their best to spread risk, at the same time generating revenue for further business expansion as well as meeting family needs. Investment in properties and unit trusts was minimal; both attracted a combined 6.4%. In short, people in various sectors of the economy were relying on their savings as an investment option, although not much was being realized in terms of interest.

6.4.3 Reliability Analysis

The exploration of the demographic details enabled the transition towards tackling of research objectives through the use of independent and dependent variables. Reliability is generally used in research to check for consistency of data instruments, especially amongst the variables. The statistic of choice is the Cronbach Alpha coefficient, which is used to confirm that the instrument's objects in the survey are altogether measuring the same constructs.

Table 6.3 shows a Cronbach's alpha coefficient of 0.733, implying that the study's instrument was consistently measuring a similar construct and hence, was reliable.

6.4.4 Establishing the Most Used Cryptocurrency in Zimbabwe

Based on Table 6.4, the most used cryptocurrency in Zimbabwe was the Bitcoin. Considering also that this is the first-generation cryptocurrency, most people had become familiar with it. The mean value (2.00) points to an agreement by respondents on the subject. The adoption of other coins was low, although Bitcoin cash, Ether and Litecoin had scored some notable points in the market.

Related to cryptocurrencies was the frequency of use of this currency by the general public, as illustrated by Table 6.5.

For those with a desire to use the cryptocurrency in foreign markets for buying and selling, this is revealed by the rate at which one acquires the coins. More than 50% of the respondents had acquired Bitcoin more than twice, as indicated by a combined value of 66.4%, although there were some who had never used a Bitcoin before. This is understandable, considering the frequency of use in Zimbabwe, with

TABLE 6.3
Consistency Data of Demographics

Cronbach's Alpha	Cronbach's Alpha Based on Standardized Items	No. of Items
0.733	0.766	36

Source: Primary data.

TABLE 6.4
Descriptive Statistics of Commonly Used Cryptocurrencies in Zimbabwe

	N	Mean	Std. Deviation
Bitcoin	63	2.00	0.000
Litecoin	63	1.13	0.338
Ripple	63	1.05	0.216
Ethereum	63	1.15	0.355
Monero	63	1.00	0.000
Eos	63	1.00	0.000
Cardano	63	1.00	0.000
Iota	63	1.00	0.000
Tron	63	1.05	0.216
Bitcoin Cash	63	1.08	0.275

Source: Primary data.

TABLE 6.5
Attainment of Cryptocurrencies in last 12 Months?

	Occurrence	Percentage	Effective percentage	Cumulative percentage
Never	20	14.3	14.3	14.3
Once	27	19.2	19.2	33.5
2–5 times	56	40.0	40.0	73.6
More than 6 times	37	26.4	26.4	100.0
Total	**140**	**100.0**	**100.0**	

Source: Primary data.

the general populace having no idea about its use. The intended use drives one to acquire the currency well in advance, especially when one plans to make overseas payments, as indicated in Table 6.6.

Table 6.6 translates the drive in acquiring of the currency in Table 6.9 to Table 6.10, where 63.6% of the respondents were expecting to use the currency soon for transactions. The main reason behind the desire for cryptocurrency is the nature of the Zimbabwe economy and the absence of local currency that has value for international purchases as well as inflation pressure in a volatile economy. A significant percentage of respondents had no intention of using cryptocurrency for purchases, as given by 28.6% of the targeted population. An independent test (Kruskal–Wallis) was performed based on the respondent's qualifications, and the statistics produced

TABLE 6.6
How Many Times Do You Expect to Use Cryptocurrencies for Cross-border Remittances (Sending or Receiving) in the Succeeding 12 Months?

	Occurrence	Percentage	Effective percentage	Cumulative percentage
Nil	40	28.6	28.6	28.6
Once	11	7.8	7.8	36.4
2–5 times	42	30	30	66.4
More than 6 times	47	33.6	33.6	100.0
Total	**140**	**98.4**	**100.0**	

Source: Primary data.

a significance value of 0.121, which is above 0.05 (threshold). The results indicate that there was no difference in opinions across the groups concerning use of cryptocurrency for the next 12 months. The mean values of these groups were statistically insignificant. The analysis is based on the Chi-squared approach, and it assumes that the data is not normally distributed.

6.4.5 Factors Influencing Cryptocurrency Usage Selection

The other statistical approach used for a list of variables is data reduction through the application of principal component analysis. This is a technique mainly used for removal of correlated variables in a data set. The uncorrelated data set is then represented by a list of components, which form the basis for further analysis if need be. The values of the components should be at least 0.5 for the variables. The variable with the highest value is then represented in a component. Table 6.7 gives a summary of the variables that might need further analysis. These are factors that reflect the usage of cryptocurrency in the country and how the authorities may keep watch on this, especially for a regulatory framework. The regulatory framework may be aligned in such a manner as to support and educate the public on these factors to increase adoption.

So, based on the few selected variables that push for extensive usage of cryptocurrency, the respondents indicated that trading in such currency has a lot of benefits. The factors include investment opportunities, with a factor analysis value (0.625) above 50% and worth looking out for. The other factors are generation of club money and international transactions, especially importing of second-hand vehicles directly from Japan. Other issues for further interrogation are lower transaction costs, speed and convenience in transaction, online payment for jobs overseas and most importantly, cross-border remittances. The last factor, high security features (0.704), implies that scrutiny is needed on how secure these transactions are, as is alluded to by technocrats in cryptography.

TABLE 6.7
Rotated Component Matrix

	Component				
	1	2	3	4	5
As an investment vehicle	0.625	0.137	0.266	0.106	0.071
Hold it as digital schemes to generate club money	0.644	0.233	0.268	0.050	−0.419
Transaction motive, e.g. buying ex-Japanese vehicles	0.122	0.268	−0.100	0.754	−0.099
For cross-border remittances	0.244	0.676	0.176	0.430	0.054
External money transfer	0.367	0.659	−0.096	0.048	0.023
To receive payment for online jobs	0.056	0.273	0.263	−0.005	0.786
Lower transaction costs	0.788	−0.224	−0.104	0.175	0.020
Speed and convenience of transacting	0.773	0.268	−0.118	−0.223	0.124
High security features	0.425	−0.040	0.704	−0.037	0.316
Potentially limitless application of blockchain technology in record keeping and database management	−0.070	0.752	0.113	−0.021	0.146

Abstraction Technique: Principal Component Analysis

Source: Primary data.

As stated by a Bitcoin technology expert:

> What has contributed to the non-adoption of cryptocurrencies I would say is the Reserve Bank, which has not been very supportive of cryptocurrencies. Secondly, it's the shortage of US dollars. Why, because to buy your crypto you need to buy it with real money on your exchanges like the bitraxs, your coin base and those people don't take the bond notes, and our Master cards don't work outside of the country. So, I believe that inhibited cryptocurrencies. What contributed to the adoption like I said the fear of missing out and secondly, I would say just the high number of people talking about the crypto. Also, the gains that Bitcoin made last year made crypto an attractive investment.

Non-adoption of cryptocurrencies has been largely due to RBZ not being supportive of cryptocurrencies and lack of the US dollar in the market to purchase cryptocurrencies, coupled with the hazards of Ponzi schemes, scams and bubble burst. Furthermore, lack of retailers accepting virtual currencies as well as regulatory warnings have just compounded the fears.

On the other hand, adoption has been catalyzed by fear of missing out, peer pressure and the high gains made by Bitcoin last year. This is attested to by a Bitcoin entrepreneur, who had the following to say:

> What is increasing the adoption of cryptocurrency are the cash crisis, large diaspora contingent, cheaper remittances, and bullish prices of cryptocurrencies. People are up against the wall and need their businesses to operate. They will look for any means and

> measures to use to make a cross-border payment. Right now, cryptocurrency is the best solution to make cross-border payments.

Further analysis is given in Table 6.8 on how exactly respondents view the selected issues.

Based on the mean values indicated in Table 6.8, respondents agreed that the motive behind cryptocurrency use is the importation of ex-Japanese vehicles. Also, cross-border remittances, receiving payment for online jobs, and its security features and absence of control from government through the RBZ provide significant beneficial factors. The mean values were all at least four (agree) based on the coding responses in SPSS. The advent of technology will always bring with it challenges and risks, and as such, trading in cryptocurrencies also presents shortcomings associated with new technology.

6.4.6 Barriers and Risks That Users Encounter in Trading Cryptocurrency in Zimbabwe

Table 6.9 gives a list of barriers and risks likely to be encountered by customers and users of the cryptocurrency financial service. The process of data extraction was also employed.

The exposition of the terminology has been done using principal analysis, that is, for selection of uncorrelated variables for further analysis. The variables highlighted for the exercise are money laundering, cyber-attacks, fear of manipulations and hacks, tax evasion, and lack of recourse to a regulator in cases of dispute. These are the factors that may fall under risks and barriers to the full adoption of cryptocurrency in Zimbabwean markets.

The Bitcoin technology expert had the following to say regarding barriers to cryptocurrency adoption:

> The number one hurdle when it comes to cryptocurrencies is the right technology. Currently we do not have the right technology to implement cryptocurrency full-scale.

TABLE 6.8
Descriptive Statistics of Applications of Cryptocurrencies

	N	Mean	Std. Deviation
Transaction motive, e.g. buying ex-Japanese vehicles	63	4.29	0.818
For cross-border remittances	63	4.16	1.244
To receive payment for online jobs	63	4.24	0.935
Lower transaction costs	63	4.15	0.938
Speed and convenience of transacting	63	4.39	0.998
High security features	63	3.82	1.274
Valid *N* (listwise)	63		

Source: Primary data.

TABLE 6.9
Rotated Component Matrix of Cryptocurrency Risk Factors

	Component		
	1	2	3
Money laundering risk, illicit economic flow into mainstream	0.671	0.205	0.092
Terrorist financing risk	0.348	0.744	−0.151
Tax evasion	−0.007	0.836	0.072
There is a possibility that someone might hack and manipulate blockchain ledger	−0.201	0.101	0.760
Prices fluctuation does not make it very good medium of exchange	0.833	−0.107	0.085
Lack of recourse to regulatory authority for dispute resolution	0.634	0.215	−0.217
Abstraction Technique: Principal Component Analysis			

Source: Primary data.

> We can do it experimentally here and there but when it comes to the full technology when it comes to mining of cryptocurrency and also when it comes to the regulation on how it's going to be monitored, it's as likely to be a very big problem so I would say those are the few challenges in the adoption. Secondly maybe we can say people are also a challenge in the sense that few people know about cryptocurrencies and are able to do transactions. So first of all, there has to be a tutorial for the whole country on how to use cryptocurrencies. Even our exchanges for example Golix once went down because of too many clients so our exchanges are not up for full scale but you being the biggest factor in the field.

This statement indicates that there is a lack of technology to implement virtual currencies on a full scale. It is also difficult to monitor and regulate cryptocurrencies. Furthermore, few people are exposed to cryptocurrencies and are able to transact with them.

The Bitcoin entrepreneur added to this puzzle by arguing as follows:

> It's easy to move money outside the country without being traceable. It poses a risk to government tax revenue if implemented whole scale because there is no way to monitor individuals' wallets and peer to peer transaction. Challenges in adoption, are lack of ownership of the Bitcoin system by an individual. Take for example when the RBZ introduces a new note, it undertakes a consumer education program to conscientize users. With cryptocurrencies no one does consumer education.

The Bitcoin exchange manager argued as follows:

> It's a disadvantage when the liquidity of the cryptocurrency is not available. So, if the country has to adopt it and not available it has to go fully cryptocurrency without other options it becomes difficult to make payments because they won't be able to get hold of the cryptocurrency. When it comes to challenges in the adoption of cryptocurrencies like I said previously on education; people need to be educated about cryptocurrencies

and how it works. Also, liquidity and availability of cryptocurrency in the markets are important to enable people to transact.

These factors were subjected to non-parametric tests to check for significant differences amongst the selected groups, which might be age or employment status or academic qualifications. A cross tabulation on the few selected barriers and risks is given in Figure 6.2.

The results in Figure 6.2 show that fear of manipulation of the system and hacking was more prevalent amongst the employed group of respondents. This is due to exposure to technology and possible threats associated with going digital in transactions. The yes column says it all, and tracking it down to the self-employed and the unemployed, the numbers decrease due to ignorance on the subject. Figure 6.3 shows employment status of the respondents against tax evasion. The issue of evasion of tax was rejected by the employed respondents due to the fact that the process is not regulated by central government, since it is a peer-to-peer transaction; hence, there is no way one is supposed to pay taxes.

The barrier of tax evasion revealed that some of the employed had no idea on the issue of cryptocurrencies. This was across the board, although the degree differed, while the self-employed and unemployed did not agree at all. The analysis of variance was performed again to check for significance of differences between academic groups on issues to do with susceptibility to cyber-financial risk. Based

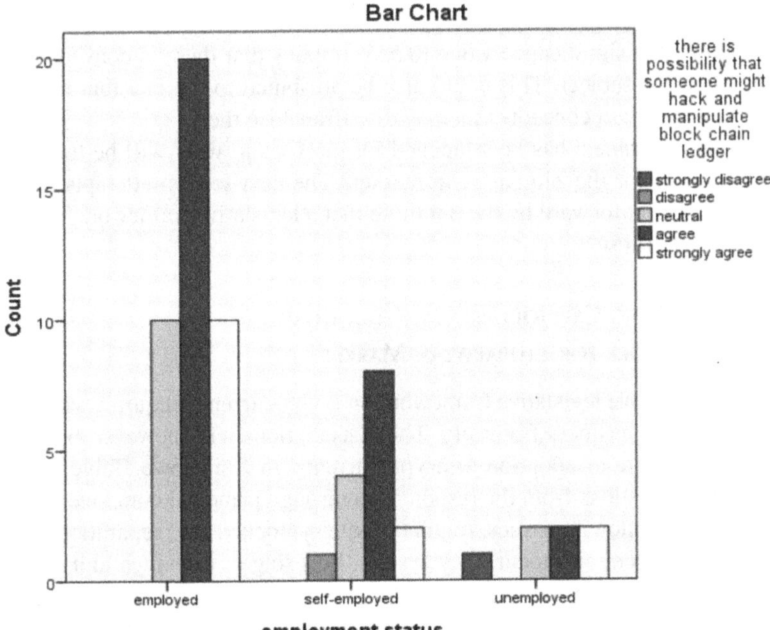

FIGURE 6.2 Distribution of confidence in blockchain ledger by employment status. (Primary data)

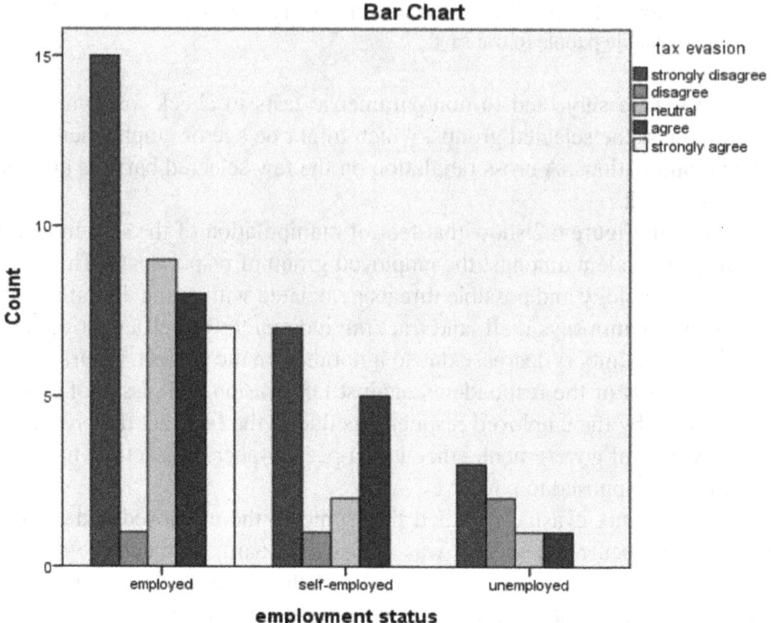

FIGURE 6.3 Distribution of tax evasion risk assessment by employment status. (Primary data)

on the results, the significance value (0.555) implies that there is convergence and agreement on the subject. This could also be attributed to the fact that individuals tend not to agree with concepts that may disenfranchise them.

Now that the subject has been opened for discussion, what will be the way forward, especially for the Zimbabwean market? The next section attempts to unlock the suggestions put forward by the public so that a legislative framework can be laid out for future purposes.

6.4.7 Proffering Cryptocurrency Regulatory Framework for Zimbabwean Market

Crafting a favourable legislative framework on cryptocurrency requires wide consultations across the business fraternity. The consultations will, however, be selective considering the rate of adoption of cryptocurrency in Zimbabwe. Table 6.10 summarizes the suggestions from the respondents through principal component analysis to bring out the following issues; regulation of cryptocurrency, separation of mainstream banking from cryptocurrency system, licensing of exchange unit in cryptocurrency, and addition of tax value to the process.

The descriptive analysis in Table 6.11 captures the views on various issues, and a mean value of 2.37 for opinions about regulation shows how informed the respondents were. This implies that they were not sure whether it should be regulated, since the idea of having a cryptocurrency was to avoid those channels of government

TABLE 6.10
Suggested Practice for Regulation (Principal Component Analysis)

	Component				
	1	2	3	4	5
In your opinion, should cryptocurrencies be regulated?	0.125	−0.076	0.138	0.857	−0.138
A law that separates the use of cryptocurrencies from mainstream banking industry	0.028	−0.095	−0.105	−0.164	0.870
All cryptocurrency exchanges to register and be licensed with RBZ for regular inspection of transactions	−0.254	0.075	0.748	0.179	−0.072
The exchange should maintain a history of transactions for users	0.213	−0.089	0.710	0.024	−0.010
Customer's fund must be kept separate from the funds of the exchange	0.020	0.845	−0.138	−0.221	−0.134
The government should add tax value on every cryptocurrency prompt made	0.782	0.013	−0.234	0.188	−0.030

Extraction Method: Principal Component Analysis

Source: Primary data.

TABLE 6.11
Descriptive Statistics for Regulation Practice

	N	Mean	Std. Deviation
In your opinion, should cryptocurrencies be regulated?	63	2.37	0.814
A law that separates the use of cryptocurrencies from mainstream banking industry	63	3.68	1.156
A law that makes cryptocurrency a legal tender (it is officially recognized as a currency that can be used to purchase)	63	4.27	0.605
All cryptocurrency exchanges to register and be licensed with RBZ for regular inspection of transactions	63	3.29	0.492
The exchange should maintain a history of transactions for users	63	3.44	0.643
Exchange should report suspicious activity to RBZ	63	4.00	0.724
Customer's fund must be kept separate from the funds of the exchange	63	4.10	0.824
The government should add tax value on every cryptocurrency prompt made	63	2.81	1.545
Jail for non-cryptocurrency users on tax evasion	63	2.81	1.524
Introduce anonymous tax on non-registered users of cryptocurrency	63	2.81	1.513
Valid *N* (listwise)	63		

Source: Primary data.

interference. This is also in line with the views of one participant, who argued as follows:

> Regulation is desirable if cryptocurrencies are to be accepted by the general population of Zimbabwe. However, it should not be so heavy handed such that it stifles innovation in the technology.
>
> **Bitcoin entrepreneur**

A similar view was expressed by another participant, who had the following argument:

> For me I would say the ability of cryptocurrencies to process cross-border payments easily is what is going to fuel their adoption. Right now, because of the regulation or regulatory ban banks are no longer opening accounts for cryptocurrency exchanges like ourselves they have closed bank accounts so regulation is very pivotal to the success of such a venture. Whether there is regulation to promote or regulation to nurture such solutions but regulation is what is needed.
>
> **Bitcoin exchange manager**

These sentiments indicate that while regulation is pivotal for the success of virtual currencies, regulation must not be so stringent as to stifle growth.

The second proposal on the list (3.68) shows some form of neutrality on the subject of enactment of laws that separate the usual banking process from the cryptocurrency procedures.

The suggestion of crypto as a legal tender might be peculiar to the Zimbabwean economy, considering where it is in the global financial market (multi-currency regime). Having a stronger currency might be the solution for Zimbabwe, because its capacity to trade in cryptocurrency is questionable, although its use will be authorized. The ban by central government on the use of cryptocurrency (RBZ, 2018) had to be reversed recently because of existing laws that do not restrict its use. Thus, in general, regulation of the cryptocurrency system is rejected by the respondents, and users are willing to take the risk associated with the platform. While there is a slow pace in the adoption of cryptocurrency, albeit due to its potential risk because of lack of regulation, one Bitcoin technology expert called for the use of blockchain technology to leverage Africa's resources for development:

> There is a very bright future in cryptocurrencies especially in Africa where we have the minds but few resources. Imagine what's going to happen if we take all natural resources and turn them into digital assets to unlock a lot of value. But because we do not have the right "know how" and the right systems this will be difficult. Other countries are trying to push for it such as Rwanda as well as Ethiopia that is using the blockchain in the farming technologies as well as South Africa that is trying to use Bitcoin in their Pick-n-Pay shops making it legal tender. All these point to cryptocurrency having a bright future.

This evidence points to the fact that the knowledge of cryptocurrency remains very limited and obscure. This finding is in line with the observation by PwC Financial

Services Institute (2015), which noted that customers will receive and accept cryptocurrency on an extensive scale only when they acquire improved knowledge and see better-quality accessibility, consistent cash exchange and inexpensive customer safety. To this end, the Congressional Research Service (CRS) 2020 further observes that numerous customers may have a deficiency of knowledge related to cryptocurrency and its working.

Criticizers of cryptocurrency have also raised apprehensions that prevailing laws and regulations do not sufficiently defend customers dealing in cryptocurrencies. Simultaneously, supporters of cryptocurrencies advise against over-regulating expertise, which will produce huge profits. Lastly, if cryptocurrency turns out to be extensively used, it could influence the ability of central banks to devise and communicate financial strategy, leading a few reviewers to claim that central banks should produce their own digital money, while others oppose this idea (CRS, 2020).

6.5 CONCLUSION AND RECOMMENDATIONS

The study revealed that the barriers and risks cryptocurrency users encounter are the susceptibility of cryptocurrency to money laundering, the possibility that someone might hack and manipulate the blockchain ledger, and also price fluctuation of the cryptocurrency, which does not make it a good medium of exchange; hence the need for regulation.

6.5.1 Barriers and Risks Encountered in Trading Using Cryptocurrencies in Zimbabwe

Two main reasons came to the fore regarding risks that affect cryptocurrency users and why some regulatory intervention may be taken to increase adoption and utilization. Firstly, there is a real risk that users face, in that their virtual currencies may be lost without any form of recourse available due to the decentralized nature of virtual currencies. A typical case is that of an exchange in the US, MtGox, which filed for bankruptcy, resulting in cryptocurrency users collectively losing $600,000 worth of bitcoins. Furthermore, there is uncertainty related to broader acceptance of Bitcoin and other cryptocurrencies because of such apparent risks. In the present unregulated environment, stakeholders use cryptocurrencies at their own risk. Central banks, including the RBZ, have generally taken the approach of warning cryptocurrency users against trading in cryptocurrencies but still fall short of implementing mitigatory measures to address these risks. Numerous individuals may be not comfortable with knowing the risks related to an unfettered cryptocurrency system. Cryptocurrency guidelines will inspire the broader acceptance of Bitcoin, which is essential for the related profits and usages of Bitcoin to be completely understood. It is, therefore, recommended that the government limit the risks and barriers encountered in trading using cryptocurrencies by implementing appropriate regulation, as elaborated in Section 6.5.2.

6.5.2 To Proffer a Cryptocurrency Regulatory Framework for Zimbabwean Markets and Users

Cryptocurrencies have three main characteristics of design and operation that present significant challenges to their regulation. These are decentralization, global presence, and diversity in current and potential users. Decentralization means that there is no central authority or control of the system, as it is a distributed network that exists on the internet. The regulation of cryptocurrencies calls for a common approach, as anyone can have access and transact from anywhere in the world based on internet connectivity, thus rendering piecemeal approaches futile. Furthermore, due to the diverse range of stakeholders in the cryptocurrency ecosystem, it would be prudent to ensure that regulation to support one player should not adversely affect the operations of another. It is almost impossible under the current system to regulate a cryptocurrency network such as Bitcoin. It is, however, possible to regulate the operations of players in the Bitcoin ecosystem. These players are dominated by exchanges, users and miners. Regulating users is again a challenge, since anyone can download a virtual wallet and use it to send and receive cryptocurrencies. It is, therefore, impossible to know who owns a wallet and how much cryptocurrency is stored in the wallet.

In light of the foregoing, it is recommended that cryptocurrencies be the main focal point in regulation. The RBZ should introduce a regulatory framework ensuring that all cryptocurrency exchanges are registered and licensed with RBZ so that they can be subjected to regular inspection. Second, it should be mandatory that all cryptocurrency exchanges maintain a history of transactions for their users. This may be achieved by applying a "know your customer" framework before opening accounts for cryptocurrency users.

Third, as part of regulations, customers' funds must be separated from the exchange's funds. Minimum capitalization requirements may be applied to act as a form of deposit protection fund.

Fourth, it should be mandatory for exchanges to be audited by an independent auditor periodically so as to conform to minimum capital requirements.

Fifth, as per its residual responsibility, the government should add value added tax on all cryptocurrency profits made; in that way, the state may benefit from the underworld of cryptocurrency and blockchain.

Sixth, there is a need to declare cryptocurrencies a digital asset and form a commission to regulate and oversee its trade.

Furthermore, there is a need to form a self-regulating body of cryptocurrency exchanges that will proffer suggestions on a code of conduct as well as coming up with standards and best practices, including guidelines for legal recourse for users.

Last but not least, RBZ should push for regional and international coordination of cryptocurrency regulation, as this cannot be a one-man show.

6.5.3 Recommendations for Future Research

Future studies are recommended on the best implementation matrix for cryptocurrency and blockchain usage in Zimbabwe. Further research needs to be done on cryptocurrency regulation with a view to a regional as well as global approach to regulation. Further research may be conducted on the aspect of initial coin offerings (ICO). While some ICO are designed to be used solely within an organization's supply chain, a number are turning out to be fraud. Most recent research related to blockchain technology has been aimed at safety and confidentiality issues. For universal usage of blockchain technology, issues like performance and latency need to be discussed in future studies.

REFERENCES

Accenture Consulting. (2017). *The (R)evolution of Money.* Blockchain Empowered Digital Currencies. https://www.accenture.com/_acnmedia/pdf-63/accenture-evolution-money-blockchain-digital-currencies.pdf

Alrousan, M.K. & Jones, E. (2016). A conceptual model of factors affecting e-commerce adoption by SME owner/managers in Jordan. International Journal of Business Information Systems, 21(3), 269–308.

Bajpai, P. (2017). "The 6 Most Important Cryptocurrencies Other Than Bitcoin." December 7, 2017 Web<https://www.investopedia.com/tech/most-important-cryptocurrencies-other-thanbitcoin/?utm_source=personalized&utm_campaign=www.investopedia.com&utm_term=11948183&utm_medium=email.

Bershidsky, L. (2017). "Bitcoin and the Value of Financial Freedom," center for the governance of change. In *Cryptocurrency and the Future of Money.* Madrid: IE University, 2019

Castillo, J. J. (2005). Challenges and alternatives to the precariousness of work and life in the current crisis (2005-2014). Ministerio de Economía y Competitividad; CSO2013-43666-R.

Center for the Governance of Change (2019). Crypto-currencies and the future of money. Going beyond the hype:How can digital currencies serve society?

Christensen et al. (2011). Disrupting class: disruptive innovation, disruptive technology, EdD, learning and teaching pedagogy.

Chun, R. (2017). *Big in Venezuela: Bitcoin Mining.* Atlantic, September 2017, at https://www.theatlantic.com/magazine/archive/2017/09/big-in-venezuela/534177/

Cooper, R.D., & Schindler, P.S. (2014). *A Companion to Qualitative Research.* London: Sage Publications.

Creswell, J.W. (2014). *Qualitative Inquiry and Research Design: Choosing Among Five Approaches* (3rd ed.). Thousand Oaks, CA: Sage.

European Central Bank (2015). *Virtual Currency Schemes: A Further Analysis.* https://www.ecb.europa.eu/pub/pdf/other/virtualcurrencyschemesen.pdf

Goepfert, J. (2018). *Worldwide Spending on Blockchain Forecast to Reach $11.7 Billion in 2022, According to New IDC Spending Guide.* July, 2018. Retrieved from IDC

Gray, R. (2006). Social, environmental and sustainability reporting and organisational value creation? Whose value? Whose creation? Accounting, Auditing and Accountability Journal. 19(6), 793-819. Emerald Group Publishing Limited. Doi 10. 1108/09513570610709872

Greener, S. (2008). *Research Methods for Business.* London: Sage Publications Company, Bonhill Street.

Helms, K. (2020). *South Africa Unveils New Cryptocurrency Rules as Usage Soars.* November 21, 2020.

Hileman, G., & Rauchs, M. (2017). *Global Cryptocurrency Benchmarking Study.* https://www.jbs.cam.ac.uk/fileadmin/user_upload/research/centres/alternativefinance/downloads/2017-global-cryptocurrency-benchmarking-study.pdf

Hulme, M. (2007). *Today's Public Relations: An Introduction.* Thousand Oaks, CA: Sage.

Jin, Y., & Kelsay, C. J. (2008). Typology and dimensionality of litigation public relations strategies: The Hewlett-Packard board pretexting scandal case. *Public Relations Review*, 34(1), 66–69.

Kazeem, Y. (2020). *The Future of Finance*. September 22.

Kumar, R. (2009). *Research Methodology: A Step-By-Step Guide for Beginners*. https://mtechlib.files.wordpress.com/2016/07/researchmethodology_stepbystepguide_kumar.pdf

Leedy, P.D., & Ormond, J.E. (2014). *Practical Research: Planning and Design*. East Lansing, MI: Pearson Education.

McKenzie, B. (2018). Blockchain and Cryptocurrency in Africa A comparative summary of the reception and regulation of Blockchain and Cryptocurrency in Africa . https://www.bakermckenzie.com

PwC Financial Services Institute (2015). *Money is no Object: Understanding the Evolving Cryptocurrency Market*. PwC, August 2015. www.pwc.com/fsi

PwC Financial Services Institute (2017). *Building Blocks: How Financial Services Can Create Trust in Blockchain*. PwC, May, 2017. www.pwc.com/fsi

PwC National Professional Services Group (2018). *Point of View: Cryptocurrencies: Time to Consider Plan B*. March. pwc.in

Rochemont, S., & Ward, O. (2019). *Understanding Central Bank Digital Currencies (CBDCs)*. Institute and Faculty of Actuaries.

Saunders, M., Lewis, P., & Thornhill, A. (2016). *Research Methods for Business Students* (7th ed.) London: Pearson Education.

Schrodt, P. (2018). Cryptocurrency will replace national currencies by 2030 according to this futurist. *Time.com*, March 1, 2018, at http://time.com/money/5178814/the-future-of-cryptocurrency/ Hereinafter Schrodt.

Sekaran, U. (2009). *Research Method* for Business A Skill Building Approach. 4th Edition, New Delhi: Wiley India.

Singleton, R. & Straits, B. C. (2018). Approaches to social research, 6th ed. Oxford University Press.

Tornatzky, L. G. and Fleischer, M. (1990). The Processes of Technological Innovation. Lexington Books.

US Congressional Research Service. (2015). *Bitcoin: Questions, Answers, and Analysis of Legal Issues*. January 28, 2015.

US Congressional Research Service Report. (2020). *Cryptocurrency: The Economics of Money and Selected Policy Issues*. Updated April 9, 2020.

Yount, W.R. (2006). *Research Design and Statistical Analysis for Christian Ministry* (4th ed.). Zimbabwe National Statistical Agency, Fourth Quarter, (2018).

7 Blockchain Technology in the Energy Industry

A Review on Policies and Regulations

Ridoan Karim and Imtiaz Sifat

CONTENTS

7.1 INTRODUCTION: BACKGROUND AND DRIVING FORCES

In recent times, energy industries all over the world have become increasingly concerned with environmental efficiency, reliability, transparency, accountability, social responsibility, social inclusiveness, and so on and so forth. These points are now more applicable than ever since COVID-19 wreaked havoc in the socio-political world order. The world has been forced to contribute to social and environmental causes and promote sustainability. Besides, many developing countries are carving

DOI: 10.1201/9781003138082-7

new strategies and policies in a bid to attract new investments, including small and medium-sized enterprise (SME) investors (Gómez-Bolaños et al., 2020).

Nevertheless, missing opportunities to attract new or foreign investors are prevailing in the energy sector of developing countries due to the challenges stemming from corruption, obscurity, technological uncertainty and political instability (Baloch et al., 2020). Furthermore, the lack of transparency in the power industry associated with demand and supply and the security of energy supply is a pressing matter for many countries, including Brazil, India, Indonesia, Bangladesh, Pakistan and many nations in Africa (Debnath & Mourshed, 2018). The conventional trends and the regulations between the government and stakeholders in the above-mentioned countries have failed to safeguard the interests of the investors. As such, considerable economic hardship could be at stake.

An economical and stable electricity supply is often central to the economic development of a nation. The advent of new disruptive technologies offers a critical environment; if employed properly, it can handle the major perils lingering in the energy sector. New innovations not only change the system technically but also change the strategies of economic development by safeguarding the environment and protecting the expectations of the stakeholders (Debnath & Mourshed, 2018).

Blockchain, as a disruptive technology, poses various opportunities and challenges to traditional business standards, including the energy industry (Oh et al., 2017). Hence, this chapter strives to delineate the opportunities and challenges of blockchain technology in the energy industry to date and analyses existing examples and cases where blockchain is utilized in the energy sector. Then, it offers an overview of the potential opportunities as well as the challenges relevant to blockchain in the energy sector. This chapter additionally sheds light on potential opportunities and challenges faced by the policy-enforcing authorities concerning the application of blockchain technology in the energy industry. The research employs the qualitative method of data collection and primarily focuses on an exploratory literature review. The analysis part is focused on broad literature sources, mainly peer-reviewed publications.

Technological development in a specific industry requires both regulatory and policy reforms. In line with that, this chapter examines the internationally functioning models of blockchain technology in the energy industry, such as innovations for trade (using "smart contracts") and investment and promoting peer-to-peer (P2P) energy production. The analysis of the models has been helpful to recommend the utilization of blockchain technology in the energy industry. This eventually paves the way toward appropriate governance frameworks to ensure decentralized, transparent and sustainable energy development. The study has two main objectives. A key contribution of this chapter is that we delve into case study-themed analysis of innovations being embraced in the energy sector: For instance, smart contracts to facilitate smoother trades. We also investigate the promotion of P2P-based production of energy. Lastly, we analyse existing policy and regulatory frameworks in place that are supposed to help the energy industry absorb the incoming innovative disruptions.

7.2 BLOCKCHAIN TECHNOLOGY IN ENERGY INDUSTRY: INTERNATIONAL USAGE AND PRACTICES

Due to the speculative bubbles in the cryptocurrency markets, modern citizens consider blockchain synonymous with applications in the finance, banking and fintech industries. In other words, the payment and mode of exchange applications of blockchain grab the largest number of headlines (Upadhyay, 2019). Nonetheless, with cryptocurrencies exhibiting enough independence in their pricing and market dynamics behaviour, the controversy over whether they are an independent asset class is nearing an end (Sifat, 2021). As such, the novelty aspect of cryptocurrencies is effectively fading. This makes possible greater propagation of news to the public about the myriad applications and extensions of blockchain-based initiatives. The nature of blockchain technology has changed the fintech industries because of its high performance and the fast and stable nature of transactions. Besides, blockchain is claimed to have the ability to enable financial trades, breaking down national currency frontiers (Chakravarty et al., 2020). Additionally, blockchain is said to reduce the burden of audits on every financial ledger (Queiroz et al., 2019).

Nevertheless, blockchain has its implications beyond digital currencies. In fact, there are various uses of this technology, ranging from land and property registration to property possession and trade protection (George et al., 2019). It can also be used, as we know, for international cash payment, fund transfer and recording intellectual property rights and copyrights for different industries (Chang et al., 2019). As for payment systems, blockchain technology is the smartest application, with greater accessibility, a more agile platform, less proneness to errors, and smoother transfers than those of the other existing payment and settlement processes. In addition to the fintech industry, blockchain has a potent future in the energy industry. For the energy market, however, blockchain can be a very expensive data storage tool (Hou et al., 2019). It is indeed unfortunate that barring the financial industry, blockchain technology has received sparse attention – particularly in the energy sector (Hald & Kinra, 2019).

Lack of understanding of this new technology has posed challenges for regulatory bodies, and many lawmakers and enforcers do not show adequate comprehension of its scope and implications (Rennock et al., 2018). Therefore, the energy market (for good reasons a heavily regulated market) will not be a suitable choice for the proliferation of decentralized blockchain technologies. Given this, it is still true that in addition to enabling money exchange, blockchain technology will be able to accommodate complex tasks such as tracking activity, property recordkeeping and exchanges, transfers of environmental and social goods, etc. As of the first quarter of 2021, another innovation of blockchain-enabled technology is gaining momentum – nonfungible tokens. These tokens are exhibiting the limitless power of blockchain technology in not only storing memory but also keeping accurate records and historical snapshots. There have already been some implementations of blockchain in the energy industry. This part of our discussion focuses on the possibilities that blockchain can bring to the energy industry.

7.2.1 Energy Trading on a P2P Basis

A recent paper by the German Energy Agency (DENA) reveals that the executives of the German energy sector opine that blockchain technology's blueprint holds promise for creating an effective, accessible and – crucially – sustainable infrastructure of power generation and distribution (Mika & Goudz, 2020). The bulk of industry practitioners also believe that eventual expansion of this sector is highly likely – as per the latest DENA survey (Burger et al., 2016). The existing views of market leaders today may be heavily influenced by internet buzz and new prospects around the consultancy sector. Due to the still embryonic nature of this technology's prospects, there are no adverse precedents of blockchain technology creating accidents or making unsavoury headlines to dissuade analysts' attitude toward a technological transition to a more sustainable energy infrastructure.

It appears that the uncertainty surrounding the way blockchain technology will eventually transform the industry is not large enough to discourage quick adoption. Indeed, several consortiums have already been formed to bring together multiple industries to further this technology's reach and accelerate its adoption. The Energy Web Foundation is expected to partner with players of the blockchain working groups in the energy sector around the world to create a flexible, open-source platform targeted directly toward the energy industry demands and design it to be energy efficient (Burger et al., 2016). Singularity (a well-known start-up in the energy industry) is in collaboration with the Rocky Mountain Institute (an American energy company) to create a partnership for the energy industry with the vision of making blockchain more productive to encourage more efficient operations in the energy sector (Bürer et al., 2019). The new partnership seeks to pursue R&D in blockchain and energy to help utilities or investors, technology creators, consumers and green energy firms realize how current market structures may be assisted, undermined or transformed.

Bloomberg New Energy Finance (BNEF) has published one case of a company using blockchain solutions to attract new consumers (McCrone et al., 2019). BNEF considered that Tokyo Electric Power Co. would like to regain customers' trust and confidence by restoring a transparent information structure in the domestic market (McCrone et al., 2019). Japan's largest power provider has developed a company called Trende, which attempted to enter solar energy production and enable P2P power purchases of solar technologies via blockchain (Martin, 2018).

Smaller companies, start-ups and pioneering firms are aggressively experimenting with P2P trading. This extends to both utility services and energy-based frameworks. Such trading through blockchain can bring possible technological changes. These changes can eventually generate more profit for some customers (Martin, 2018). Consequently, blockchain may shape the prospective future of P2P power trading, since the energy start-ups are looking forward to offering competitive deals through innovative technologies. Nevertheless, it is yet to be understood how such peer-to-peer transactions can be coordinated on a scalable level; therefore, maintaining improvements in the infrastructure that facilitates such power trading is quintessential. Several countries are already experimenting with P2P on a limited scale, adopting a pilot approach.

7.2.2 Managing Energy Supply and Demand in Real Time

Electricity markets are constantly forced to balance supply and demand. Such a balance is a challenge for both renewable and fossil fuel-based energy generation. In the transition process from fossil fuels to renewable sources, many countries are in desperate need of new flexible systems to accurately forecast the power demand and match the supply accordingly to the system. Hence, a transparent data structure (backed by blockchain technology) is also essential to evaluate energy sources and prepare them for quick adaptation in times of scarcity.

Blockchain (theoretically) is a technology that can offer stable, real-time data transmission and allow effective monitoring and maintenance of electricity-industry infrastructures. The technology can also provide quicker response times (again, in theory) in an emergency (Siemon et al., 2020). Data is expected to be protected and accessed by each stakeholder. Blockchain is anticipated to add a security and collaboration layer to the existing digital pilots by allowing quick, reliable data collection and connectivity between equipment vendors, infrastructure maintenance, and emergency response teams. In order to serve that purpose, Tennet TSO GmbH (Germany) is now partnering with battery manufacturers IBM and Sonnen GmbH to establish a non-physical line of transmission utilizing blockchain-based technology to use the surplus electricity extracted from wind turbines and store the power using batteries. The storage power is then distributed through electric grid lines in the southern part of the country (Höhne & Tiberius, 2020). This decreases the power and infrastructure cost associated with installing new lines (Höhne & Tiberius, 2020). The UK-based corporation Electron is also planning a new breakthrough in terms of the supply chain model by utilizing blockchain technology (Goranovic et al., 2017). The company is also using blockchain to build (via versatility marketplace) a framework to maintain the power demand and supply. This phenomenon is dubbed Energy eBay because of its tremendous prospects and role in stimulating widespread industry participation (Goranovic et al., 2017).

Developing countries, still undergoing the ultimate hurdle of ensuring sufficient energy supply, can implement blockchain-powered applications to make the demand forecasting strong and accurate via real-time data analytics from the system. Such exposure of information can help the regulators to effectively manage the supply of electricity following its economic mandates. Real-time demand signals emitting from a blockchain-based ledger include the ability to transact between all the assets demanded and supplied, therefore helping to make effective decisions (Dong et al., 2018).

7.2.3 Promoting Investments in Energy Industries

According to Crunchbase (a multinational start-up database), at this moment, 140 companies are rapidly heading toward the blockchain-empowered energy sector (Andoni et al., 2019). Start-ups appear to be leading advancement in transforming the business models in the electricity industry, riding on the back of blockchain-based applications (Deign, 2017). Blockchain technology offers transparency to investors.

Governments are aiming to form strategic alliances with the energy start-ups with a view to employing the blockchain technology for quicker responses and incorporating trust in the system to break the conventional model of the slow licensing process (Andoni et al., 2019).

Start-ups can act as a potent player, transforming the energy market through blockchain-enabled technologies. Nonetheless, to be a successful disruptive actor for the energy revolution in reality, there is a crying need for solemn commitments and support from governments. Risks and incentives to achieve priorities such as climate change and energy transformation targets must be carefully considered when working with the appropriate grid infrastructure. Blockchain technology can trace the energy investments in different electricity sources and promote sustainable energy solutions.

US-based company TransActive Grid, Power Ledger and Singularity from Australia, Ideo CoLab, etc. are some of the start-ups in the energy industries that are efficiently utilizing blockchain technology (Wang et al., 2017). These companies have demonstrated that specific usage of blockchain technology increases the efficiency in supply analyses of the energy industry, helps in pricing and project profits, assists in electricity sharing, provides options for automated power plants, and performs microgrid management (Agency, 2016). In the meantime, such innovations can also be exploited to trade renewable energy certificates (RECs), which are typically given to the producers of solar energy based on the anticipated production demands rather than actual figures. Several innovative ventures are already engaged in designing grids and systems to facilitate power production in order to be able to get these certificates. One such firm, Volt Markets, supplies energy and monitors a trade network built atop smart contracts underpinned by Ethereum's technology (Henderson et al., 2018).

7.2.4 Increasing Energy Access in Developing Countries

In the energy sector, blockchain can also retain its original purpose and be used as the cryptocurrency for monetary transactions; in fact, some companies have already begun this campaign: For instance, SolarCoin, Bankymoon and BlockCharge, to name a few (Kumar, 2018). In terms of blockchain usage, start-ups are still involved, while applications are catching up. Others would go for joint ventures and cooperations as a viable option. Hence, the value blockchain proposes would be the removal of brokers as intermediaries between parties.

In principle, a complete decentralization of the energy sector through blockchain implementation may be accomplished if financial transactions are removed from central control. These advances may further allow room for inventions that over time would increase access to electricity in developing countries. Take smart prepaid meters as an example: A technology that only releases power to residential consumers after they have updated their accounts and moved money to their energy supplier (Chitchyan & Murkin, 2018).

This program could be advantageous for high-inflation countries like the BRICS economies or South Asian nations. Conceptualizations of the technology came from a South African enterprise – Bankymoon – which leverages bitcoin's network to enable remote payment systems (Dogo et al., 2019). For poor economies that suffer from a lack of economic resources, such innovations bode well. As for the altruists who want to donate to help schools carry on their activities, they can bypass the traditional restrictions and dive right into crypto-based systems to contribute to the school's smart meter. Consequently, the schools will receive power credit and instantaneously achieve energy self-sufficiency without unnecessary hassles from intermediaries (Henderson et al., 2018).

Blockchain can also enable crowdfunding in the energy sector and increase energy access in developing countries (Arnold et al., 2019). Numerous solar panel schemes are still left unfunded in many parts of Africa, most of which are worth less than 1 million USD (Brilliantova & Thurner, 2019). Crowd-financing could fill in this funding gap; people around the world would be able to buy the photovoltaic cells that will make up the solar panels on African homes (Higgins, 2016). The solar panels are only installed when enough solar cells are pre-purchased. Throughout the crowd-sale duration (a certain finite amount of days), the total an investor wants to pay is assigned to the number of solar cells he can get for it (because of bitcoin's volatility) (Brilliantova & Thurner, 2019). These cells provide African households with electricity, and households, in turn, pay investors a rental income in bitcoin for several years. The blockchain platform is used in this case to fund access to electricity.

7.2.5 Regulatory Reporting and Compliance

Regulators constantly expect energy and resource firms to have large volumes of data that can be evaluated to identify non-compliance with legal and regulatory requirements (Diestelmeier, 2017). Collecting and analyzing the necessary data is a big challenge for the latest technology and applications. There is also a substantial danger of the data falling into the wrong hands and being misused, exposing confidential business information and placing a company at a strategic disadvantage.

Blockchain could theoretically solve many of these problems, facilitating accountability and encouraging regulators to access secure, transparent data safely at source and encouraging businesses to keep tight control over what information is accessible and who can access it (Diestelmeier, 2017). A significant side advantage of having such a forum to exchange knowledge with regulators is that it will establish an industry-standard data format, which is impossible at present (Bürer et al., 2019).

Energy companies are particularly worried about trade secrets. Private blockchain networks offer pre-approved parties' authorization for data and limited consortium entry. Private and cooperative blockchains offer an intermediate alternative before the required privacy features of business demand can be introduced by the public blockchains (Bürer et al., 2019). Blockchain's core points of focus on the electricity market are cost reduction, sustainable development and increased accountability without sacrificing privacy, which can bring a paradigm shift to the whole energy sector of a developing country within a short span.

7.3 PROSPECTS OF BLOCKCHAIN TECHNOLOGY IN ENERGY INDUSTRY

Blockchain technologies are deemed to be one of the most forthcoming pathways to expedite the entrance of various energy sources. With plausible reasons, structuring a blockchain-based communication system certainly offers appealing characteristics. Blockchain can adequately respond to numerous obstructions found in the energy sector of developing countries. Furthermore, it can cater to the corresponding requirements of the local trade in electricity production and consumption from a particular vantage point to provide the replenish the shortage amount to the power grid.

A wide range of business and operations relating to the energy industry may be regulated with blockchain technology (Edeland & Mörk, 2018). Smart metering and smart contracts generated from blockchain technology can usher in a new era of automated billing for customers and distributed generators. Pay-as-you-go platforms can benefit both consumers and utility companies. The potential of energy micropayments can only be exploited with the help of blockchain technology (Burger et al., 2016).

Energy sales practices will change in most countries accordingly, depending on the environmental aspects, individual preferences and the customer's profile (Burger et al., 2016). Blockchains can classify market energy trends with artificial intelligence (AI) techniques like machine learning (ML) (Singh et al., 2020). Distributed trading platforms allowed by blockchain could dislocate business operations such as risk management, demand control and commodity trading (Andoni et al., 2019).

Blockchains can also expand the power of autonomous microgrid and energy systems (Burger et al., 2016). Blockchains can theoretically be used to provide smart grid solutions with the combined usage of smart meters, automated sensors, network security equipment, energy storage and control systems, and smart home energy controllers and building control systems (Khaqqi et al., 2018). These intelligent systems can be used with enhanced potentiality through the application of blockchain technology-enabled smart grids. Besides, blockchains can assist with network security, flexibility or asset management of decentralized networks and can offer integrated flexible trading networks to leverage scalable infrastructure (Pan et al., 2020). Smart contracts will also theoretically ease energy trading and enhance energy mobility, potentially lowering energy tariffs. Immutable databases and consistent procedures can dramatically improve auditing and compliance with regulations (Thomas et al., 2019).

Blockchain in the energy industry can attract prospective investments. Tyro firms or inexperienced start-ups can enjoy low barriers of entry into this sector via P2P energy trading models. For a developing economy, blockchain certainly has the answers to numerous challenges regarding the decentralized energy production scenario (Giotitsas et al., 2015). It can further aid the policymakers of the country to discern the energy demand and deviations in real time through grid simulations. For policymakers, blockchain is also helpful in demand planning and all the accompanying considerations through the implementation of this technology (Kiviat, 2015). As a result, the costs of electricity would be open to all parties concerned. This includes the grid responsible for distributing power and utility companies.

Another important benefit of implementing blockchain in the energy industry is that all the electricity that the network delivers can be allocated in small units to individual customers. This allows the precise calculation of all electricity generated and consumed. A much-improved database would help to finalize network operations better at distribution and transmission levels. Figures 7.1 and 7.2 present the current market roles and how the system would change if blockchain technology were used.

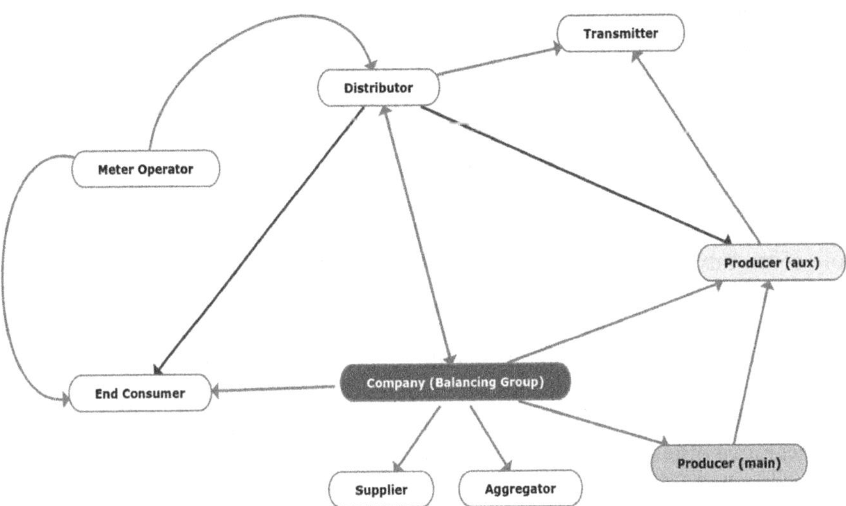

FIGURE 7.1 Contemporary energy market.

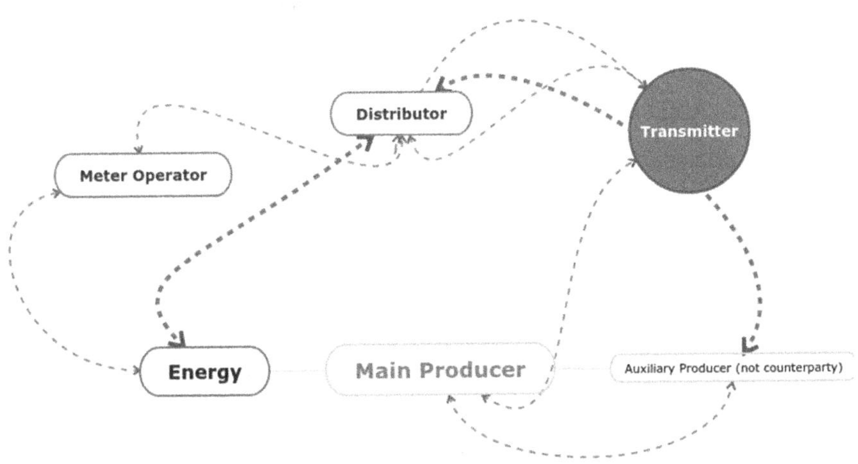

FIGURE 7.2 Energy market structure using blockchain.

Blockchain technology enables direct contracts between energy consumers and power producers. Therefore, all consumers must comply with energy laws, especially the laws relating to energy safety and risk management. The function of meter operators will change; they will no longer need to gather and monitor data themselves, since all information on use and transaction is immediately and precisely shared using the blockchain technology through smart contracts.

To deliver electricity directly to consumers from a power supplier and complete a financial exchange between the parties on blockchain technology, both the parties must integrate within the system. Since the energy consumer (see Figures 7.1 and 7.2) will continue to use power from the network, the consumer will eventually be the meter operator. The practical challenge begins as the question remains whether the consumer can effectively understand and adhere to the technicalities relating to meter reading (PricewaterhouseCoopers, 2016). Blockchains are still to be implemented in energy market communication systems. A blockchain model in the energy industry can create a substantial financial and organizational obstacle. One significant barrier stopping the implementation of transaction models based on blockchain is that they must conform to existing regulatory requirements (Diestelmeier, 2019). Any of the advantages that a decentralized P2P relationship structure would offer are then missed.

A blockchain roll-out will have a significant effect on energy market competitiveness. There is a possibility that small or local firms will face fewer obstacles, while big investors will be financially affected. Conversely, many experts believe there is a solution to this (Löbbe & Hackbarth, 2017). Setting up private blockchains with lucrative offers for consumers will allow them to remain in the market (Löbbe & Hackbarth, 2017). In such a case, small vendors will fail to include them in the transaction model of the energy market. The existing insecurity regarding legal and regulatory approval is another hurdle for blockchain implementation in the energy industry. Compared with the anticipated blockchain-based energy model, the existing regulatory structures are focused on a simple distribution of corporate and legal obligations.

7.4 LEGAL AND REGULATORY CHALLENGES

With every dramatic transition to new technologies, blockchain poses a range of legal concerns. Legal concerns vary from private to public international law and from financial to energy laws and regulations, based on the nature and scope of blockchain applications. Some of the legal concerns of blockchain relating to energy industries are highlighted here.

7.4.1 Applicable Law

The topic relating to applicable law is the first issue posed in blockchain. In blockchain, jurisdiction is important since the parties of each transaction will be in separate countries. Hence, determining the applicable law will be challenging. For each transnational transaction in blockchain, all the parties to the smart contract will feel the necessity to understand applicable/governing law. This problem only does not arise in situations where the parties to the agreement reside within the EU, as the legislation applicable to contractual obligations remains the same: The "Regulation of the European Parliament and of the Council on the law applicable to contractual obligations (Rome I58)" (Giancaspro, 2017). But what happens if the contracting parties are not in the EU?

The matter should then be dealt with under either private international law or the applicable law mentioned in the contract (Governatori et al., 2018). A possible solution to this may be providing a range of alternative options to choose the applicable law in a smart contract by clicking on the preferred country or region. In the case of imposing blockchain applications in the energy industry, selecting a proper governing law for each transaction may lead to complex legal and regulatory situations, which the tech giants and the government should consider with a more profound and holistic approach.

7.4.2 Identity within a Blockchain

Most of the energy-related blockchain applications so far are held in public blockchains. All transactions in public blockchains are easily accessible. Nevertheless, transactions in public blockchains do not indicate that the identity of the parties is always understandable. For example, the parties in a bitcoin transaction can participate anonymously. This can be an added advantage for cryptocurrency situations; nevertheless, it contributes to difficulties for energy industries, both politically and philosophically (Diestelmeier, 2019). Unidentifiability will lead to a situation where the stakeholders cannot identify the real problem or hold any authority responsible for a wrong, which stands against the core philosophy of introducing blockchain into the energy industry.

In this case, money laundering on Emission Trading System (ETS) markets, or market theft due to confidentiality, will remain as a persisting challenge in the energy market even though blockchain applications are introduced. Therefore, the existing situation of market exploitation and uneven competition by large energy vendors and suppliers cannot be easily changed through blockchain applications (Maksimenko, 2019). This will affect the integrity of the relevant traditional economy unless blockchain applications that can identify the parties in energy transactions are implemented. Therefore, a range of regulatory reforms have been developed relating to blockchain applications, especially in the financial sector. While there are expert arguments on the blockchain recognition/identification of electronic profiles (Caytas, 2017), such access to identity remains unanswered due to technical and regulatory barriers.

7.4.3 Liability and Responsibility

The perfect blockchain paradigm allows for a mechanism that totally operates without a responsible central authority (Radziwill, 2018). This means that a legal personality is not defined in blockchain applications (Kshetri, 2017). If any damages arise, the existing liability laws and regimes are not sufficient to deal with the compensation (Corrales Compagnucci et al., 2020). This constitutes a series of legal dilemmas. Hence, specific liability laws relating to blockchain in energy industries are needed to guarantee the legal responsibilities of the parties involved. Specific laws are needed to administer the liability principles in payment defaults, technological breakdowns, deliberate non-performance, etc. As the energy industry typically entails using vital infrastructure, the stakeholders need an emergency plan to specify the protocols in case of a disaster.

7.4.4 Personal Data

One of the basic features of blockchain contradicts the personal data laws. Most of the laws relating to data protection state that personal data must be deleted after they have served their purpose. Industry experts and the scholarly community have long pointed out that blockchain technology engenders privacy concerns. Prominent among them is the contravention of the EU's data protection regulation, popularly called GPDR (Feng et al., 2019). The original intent of this technology was to facilitate P2P transactions without a centralized data verifier. Therefore, no single party is supposed to hold the power to hold a particular network hostage. For this reason, the entire system should be permissionless so that everyone in this system can equally access the data within. In the case of sensitive data in the healthcare industry, blockchain certainly breaches several privacy principles, such as "Rights of Data Subjects", "Security Principle", "Retention Principle" and "Right to be Forgotten" (Fabiano, 2018).

7.4.5 Challenges with the Financial Market Regulation

As finance transactions move from energy providers or banks to a P2P system, the question arises of who is accountable for securing each financial transaction's settlement. Such a duty of care could not be levied on energy users and providers alone. Instead, an individual body or a network operator should be formed to conform to the requirements of financial services in compliance with the banking laws and regulations.

7.5 POLICY RECOMMENDATIONS

Policymakers need to collaborate with the industry and ensure regulatory enforcement to benefit from blockchain technologies. In some instances, policymakers cannot implement comprehensive, prescriptive and detailed guidelines, as the technology itself is in its embryo stage. Policymakers must promote technology through flexible laws to cover a large spectrum of technical applications. In all cases, regulators should collaborate with stakeholders to ensure that the implemented laws/rules and regulations are well thought out for the industry.

Among the proposals put forward as a remedy, regulatory sandboxes stand out as a highly promising solution pertaining to distributed ledger technology (Ahl et al., 2020). These allow an opportunity for program testing in isolated and highly regulated environments with real people and bona fide consumers. The entire process is also free from threats of regulatory intervention or retaliation. Furthermore, policymakers may have an opportunity to understand better the underlying technology in action and work closely with industry stakeholders to further refine and fine-tune the enforcement procedure. The result here will be the formulation of a regulatory framework that encourages and nurtures innovation without compromising security, privacy, accountability and forward-thinking solutions (Yeoh, 2017).

One challenge of regulatory sandboxes is to ensure that commercial energy sectors are appealing to them. Sandboxes should be promoting innovation and enabling energy start-ups to expand rather than merely offering legal value. The proliferation of decentralized systems affords regulators discretionary powers that can be misused or misapplied. For instance, participants may inadvertently fall into liability of compliance breach traps. Delicate handling of these matters is complicated by a lack of precedents in the legal materials. To further illustrate this point, let us consider privacy issues in sensitive systems. A few years back, the French government adopted a position that only participants actively injecting data into a system can be treated as bona fide data miners. This discounts the roles of minders or nodes who play a key role in transaction verification. Whether such positions will truly safeguard public interest in trustless systems is a matter of speculation. Whether or how transparency and accountability will be ensured for privacy and personal data spaces is another major concern.

The governance of any blockchain system is a critical issue, which is often overlooked. How a blockchain is governed defines its success and adoption in many ways. Interestingly, different aspects of a blockchain governance are analogous to how an organization or a consortium of organizations distributes responsibilities among themselves. Each organization/industry must understand this before they decide to adopt blockchain for their solutions. This calls for a new governance model, which itself might be challenging.

A regulatory framework surrounding any novel technology instils confidence among the adopters. On the contrary, the lack of any such framework creates uncertainties. Therefore, a blockchain-friendly regulatory framework can be an effective tool to overcome many challenges within an industry. It is evident that law and enforcement would work as a significant determinant to either open blockchains in the energy market or close them for good. The fast development of proactive regulation remains a challenge to the regulators. Devising this environment would ensure consumer protection and guarantee a secure and reliable supply of electricity. An absence of law or regulation regarding blockchain just might make investors sceptical about funding. Before investing in new technologies, it is common for the energy industry to be patient and wait for a long time for technology to catch up; the same goes for the emergence of policies too.

7.6 CONCLUSION

Recent policy actions concerning blockchain technologies are mostly concentrated into digital currencies and financial applications. Nevertheless, breaking the boundaries of fintech, blockchain has so much to promise in the energy industry. Though the digital currency is illegal in many countries, the uses of blockchain technology are completely legal. It is imperative to use the potential of every new technology to achieve maximal growth and welfare for all countries. Blockchain has the power to unite various fragmented systems in order to generate insights and assess the value of care properly. In the long haul, a nationwide blockchain network to maintain energy records may enhance efficiency and bolster better investments. However, blockchain technology has not reached maturity, nor is it a silver bullet or an instant solution. Several organizational, technical and behavioural economic challenges need to be surmounted before adopting blockchain in the energy sector. Therefore, though this technology has tremendous potential, it is very important to note that such a technology should be dealt with cautiously and with sincerity.

REFERENCES

Agency, D. (2016). *Blockchain in the energy transition. A survey among decision-makers in the German energy industry.* https://www.dena.de/fileadmin/dena/Dokumente/Meldungen/dena_ESMT_ Studie_blockchain_englisch.pdf

Ahl, A., Yarime, M., Goto, M., Chopra, S.S., Kumar, N.M., Tanaka, K., & Sagawa, D. (2020). Exploring blockchain for the energy transition: Opportunities and challenges based on a case study in Japan. *Renewable and Sustainable Energy Reviews, 117*, 109488. https://doi.org/10.1016/j.rser.2019.109488

Andoni, M., Robu, V., Flynn, D., Abram, S., Geach, D., Jenkins, D., McCallum, P., & Peacock, A. (2019). Blockchain technology in the energy sector: A systematic review of challenges and opportunities. *Renewable and Sustainable Energy Reviews, 100*, 143–174. https://doi.org/10.1016/j.rser.2018.10.014

Arnold, L., Brennecke, M., Camus, P., Fridgen, G., Guggenberger, T., Radszuwill, S., Rieger, A., Schweizer, A., & Urbach, N. (2019). Blockchain and initial coin offerings: Blockchain's implications for crowdfunding. In *Business Transformation through Blockchain.* (pp. 233–272). Springer International Publishing. https://doi.org/10.1007/978-3-319-98911-2_8

Baloch, Z.A., Tan, Q., Iqbal, N., Mohsin, M., Abbas, Q., Iqbal, W., & Chaudhry, I.S. (2020). Trilemma assessment of energy intensity, efficiency, and environmental index: Evidence from BRICS countries. *Environmental Science and Pollution Research 27*(27), 34337–34347.

Brilliantova, V., & Thurner, T.W. (2019). Blockchain and the future of energy. *Technology in Society, 57*, 38–45. https://doi.org/10.1016/j.techsoc.2018.11.001

Bürer, M.J., de Lapparent, M., Pallotta, V., Capezzali, M., & Carpita, M. (2019). Use cases for blockchain in the energy industry opportunities of emerging business models and related risks. *Computers and Industrial Engineering, 137*, 106002. https://doi.org/10.1016/j.cie.2019.106002

Burger, C., Kuhlmann, A., Richard, P.R., & Weinmann, J. (2016). *Blockchain in the energy transition. A survey among decision-makers in the German energy industry.* German Energy Agency.

Caytas, J. (2017). Blockchain in the US regulatory setting: Evidentiary use in vermont, delaware, and elsewhere. *The Columbia Science and Technology Law Review.*

Chakravarty, S.R., Sarkar, P., Chakravarty, S.R., & Sarkar, P. (2020). Applications of blockchain. In *An Introduction to Algorithmic Finance, Algorithmic Trading and Blockchain*, (pp. 177–179). Emerald Publishing Limited.https://doi.org/10.1108/978-1-78973-893-320201021

Chang, S.E., Chen, Y.C., & Wu, T.C. (2019). Exploring blockchain technology in international trade: Business process re-engineering for letter of credit. *Industrial Management and Data Systems*. https://doi.org/10.1108/IMDS-12-2018-0568

Chitchyan, R., & Murkin, J. (2018). Review of blockchain technology and its expectations: Case of the energy sector. *arXiv:1803.03567.*

Corrales Compagnucci, M., Kono, T., & Teramoto, S. (2020). Legal aspects of decentralized and platform-driven economies. In Legal Tech and the New Sharing Economy, (pp. 1–11). Springer Singapore. https://doi.org/10.1007/978-981-15-1350-3_1

Debnath, K.B., & Mourshed, M. (2018). Challenges and gaps for energy planning models in the developing-world context. *Nature Energy*, *3*(3), 172–184.

Deign, J. (2017). *15 Firms leading the way on energy blockchain*. Greentech Media.

Diestelmeier, L. (2017). Regulating for blockchain technology in the electricity sector: Sharing electricity-and opening Pandora's Box? In 16th Annual Conference in Science, Technology, and Society Studies. Kyoto, Japan

Diestelmeier, L. (2019). Changing power: Shifting the role of electricity consumers with blockchain technology: Policy implications for EU electricity law. *Energy Policy*, *128*, 189–196. https://doi.org/10.1016/j.enpol.2018.12.065

Dogo, E.M., Salami, A.F., Nwulu, N.I., & Aigbavboa, C.O. (2019). *Blockchain and internet of things-based technologies for intelligent water management system*. Artificial Intelligence in IoT (pp. 129–150). Springer International Publishing. https://doi.org/10.1007/978-3-030-04110-6_7

Dong, Z., Luo, F., & Liang, G. (2018). Blockchain: A secure, decentralized, trusted cyber infrastructure solution for future energy systems. *Journal of Modern Power Systems and Clean Energy*, *6*(5), 958–967. https://doi.org/10.1007/s40565-018-0418-0

Edeland, C., & Mörk, T. (2018). Blockchain technology in the energy transition an exploratory study on how electric utilities blockchain technology in the energy transition. *Trita-Itm-Ex Nv*, *2018*, 78.

Fabiano, N. (2018). The Internet of Things ecosystem: The blockchain and data protection issues. *Advances in Science, Technology and Engineering Systems*, *3*, (2), 01–07. https://doi.org/10.25046/aj030201

Feng, Q., He, D., Zeadally, S., Khan, M.K., & Kumar, N. (2019). A survey on privacy protection in blockchain system. *Journal of Network and Computer Applications*. https://doi.org/10.1016/j.jnca.2018.10.020

George, R.P., Peterson, B.L., Yaros, O., Beam, D.L., Dibbell, J.M., & Moore, R.C. (2019). Blockchain for business. *Journal of Investment Compliance*, *20*(1), 17–21https://doi.org/10.1108/joic-01-2019-0001

Giancaspro, M. (2017). Is a 'smart contract' really a smart idea? Insights from a legal perspective. *Computer Law and Security Review*, *33*(6), 825–835. https://doi.org/10.1016/j.clsr.2017.05.007

Giotitsas, C., Pazaitis, A., & Kostakis, V. (2015). A peer-to-peer approach to energy production. *Technology in Society*, *42*, 28–38). https://doi.org/10.1016/j.techsoc.2015.02.002

Gómez-Bolaños, E., Hurtado-Torres, N.E., & Delgado-Márquez, B.L. (2020). Disentangling the influence of internationalization on sustainability development: Evidence from the energy sector. *Business Strategy and the Environment*. *29*(1), 229–239.

Goranovic, A., Meisel, M., Fotiadis, L., Wilker, S., Treytl, A., & Sauter, T. (2017). Blockchain applications in microgrids: An overview of current projects and concepts. In Proceedings IECON 2017: 43rd Annual Conference of the IEEE Industrial Electronics Society. Beijing, China. https://doi.org/10.1109/IECON.2017.8217069

Governatori, G., Idelberger, F., Milosevic, Z., Riveret, R., Sartor, G., & Xu, X. (2018). On legal contracts, imperative and declarative smart contracts, and blockchain systems. *Artificial Intelligence and Law*, *26*(4), 377–409. https://doi.org/10.1007/s10506-018-9223-3

Hald, K.S., & Kinra, A. (2019). How the blockchain enables and constrains supply chain performance. *International Journal of Physical Distribution and Logistics Management, 49*(4), 376–397. https://doi.org/10.1108/IJPDLM-02-2019-0063

Henderson, K., Rogers, M., & Knoll, E. (2018). *What every utility CEO should know about blockchain*. McKinsey & Company.

Higgins, S. (2016). *How bitcoin brought electricity to a South African school*. CoinDesk.

Höhne, S., & Tiberius, V. (2020). Powered by blockchain: Forecasting blockchain use in the electricity market. *International Journal of Energy Sector Management. 14*(6), 1221–1238 https://doi.org/10.1108/IJESM-10-2019-0002

Hou, W., Guo, L., & Ning, Z. (2019). Local electricity storage for blockchain-based energy trading in industrial internet of things. *IEEE Transactions on Industrial Informatics, 15*(6), 3610–3619. https://doi.org/10.1109/TII.2019.2900401

Khaqqi, K.N., Sikorski, J.J., Hadinoto, K., & Kraft, M. (2018). Incorporating seller/buyer reputation-based system in blockchain-enabled emission trading application. *Applied Energy, 209*, 8–19. https://doi.org/10.1016/j.apenergy.2017.10.070

Kiviat, T.I. (2015). Beyond bitcoin: Issues in regulating blockchain transactions. *Duke Law Journal.*

Kshetri, N. (2017). Blockchain's roles in strengthening cybersecurity and protecting privacy. *Telecommunications Policy, 41*, 1027–1038. https://doi.org/10.1016/j.telpol.2017.09.003

Kumar, N.M. (2018). Blockchain: Enabling wide range of services in distributed energy system. *Beni-Suef University Journal of Basic and Applied Sciences, 7*(4), 701–704. https://doi.org/10.1016/j.bjbas.2018.08.003

Löbbe, S., & Hackbarth, A. (2017). The transformation of the German electricity sector and the emergence of new business models in distributed energy systems. In *Innovation and Disruption at the Grid's Edge*.(pp. 287–318). Elsevier. https://doi.org/10.1016/B978-0-12-811758-3.00015-2

Maksimenko, P.N. (2019). Legal risks and opportunities related to the use of blockchain technology in the energy sector. Energy Law Forum, *1*, 52–58. https://doi.org/10.18572/2312-4350-2019-1-52-58

Martin, C. (2018). *How blockchain is threatening to kill the traditional utility*. Bloomberg. Com.

McCrone, A., Ajadi, T., Boyle, R., Strahan, D., Kimmel, M., Collins, B., Cheung, A., & Becker, L. (2019). Global trends in renewable energy investment 2019. *Bloomberg New Energy Finance.*

Mika, B., & Goudz, A. (2020). Blockchain-technology in the energy industry: Blockchain as a driver of the energy revolution? With focus on the situation in Germany. *Energy Systems, 12*(2), 285–355. https://doi.org/10.1007/s12667-020-00391-y

Oh, S.-C., Kim, M.-S., Park, Y., Roh, G.-T., & Lee, C.-W. (2017). Implementation of blockchain-based energy trading system. *Asia Pacific Journal of Innovation and Entrepreneurship*. https://doi.org/10.1108/apjie-12-2017-037

Pan, X., Song, M., Ai, B., & Ming, Y. (2020). Blockchain technology and enterprise operational capabilities: An empirical test. *International Journal of Information Management. 52*, 101946. https://doi.org/10.1016/j.ijinfomgt.2019.05.002

PricewaterhouseCoopers. (2016). Blockchain: An opportunity for energy producers and consumers? In *Pwc.Com.*

Queiroz, M.M., Telles, R., & Bonilla, S.H. (2019). Blockchain and supply chain management integration: A systematic review of the literature. *Supply Chain Management: An International Journal, 25*(2), 241–254. https://doi.org/10.1108/SCM-03-2018-0143

Radziwill, N. (2018). Blockchain revolution: How the technology behind bitcoin is changing money, business, and the world. *Quality Management Journal, 25*, (1), 64–65. https://doi.org/10.1080/10686967.2018.1404373

Rennock, M.J.W., Cohn, A., & Butcher, J.R. (2018). Blockchain technology regulatory and investigations. *The Journal Litigation.*

Siemon, C., Rueckel, D., & Krumay, B. (2020). Blockchain technology for emergency response. In Proceedings of the 53rd Hawaii International Conference on System Sciences. Hawaii, U.S. https://doi.org/10.24251/hicss.2020.075

Sifat, I. (2021). On cryptocurrencies as an independent asset class: Long-horizon and COVID-19 pandemic era decoupling from global sentiments. *Finance Research Letters, 43,* 102013. https://doi.org/10.1016/j.frl.2021.102013

Singh, S.K., Rathore, S., & Park, J.H. (2020). BlockIoTIntelligence: A blockchain-enabled intelligent IoT architecture with artificial intelligence. *Future Generation Computer Systems, 110,* 721–743. https://doi.org/10.1016/j.future.2019.09.002.

Thomas, L., Zhou, Y., Long, C., Wu, J., & Jenkins, N. (2019). A general form of smart contract for decentralized energy systems management. *Nature Energy, 4*(2), 140–149. https://doi.org/10.1038/s41560-018-0317-7

Upadhyay, N. (2019). Transforming social media business models through blockchain. In *Transforming Social Media Business Models Through Blockchain* (pp. 1–13). Emerald Publishers.https://doi.org/10.1108/9781838672997

Wang, J., Wang, Q., Zhou, N., & Chi, Y. (2017). A novel electricity transaction mode of microgrids based on blockchain and continuous double auction. *Energies, 10*(12), 1971. https://doi.org/10.3390/en10121971

Yeoh, P. (2017). Regulatory issues in blockchain technology. *Journal of Financial Regulation and Compliance, 25,* (2), 196–208 https://doi.org/10.1108/JFRC-08-2016-0068

8 Exploring Blockchain-based Government Services in India

Gayatri Doctor and Kratika Narain

CONTENTS

DOI: 10.1201/9781003138082-8

8.1 INTRODUCTION: BACKGROUND AND DRIVING FORCES

Today's world mainly relies on information that is generated and distributed among the public on a large scale, but unfortunately, the authenticity of the information is always in question. Recently, a new technological breakthrough called blockchain has redefined the enterprise of "trust", making the information business integrally trustworthy. It is considered especially important for government agencies to have complete, secure, accurate and reliable information exchange in various fields. Organizations around the world can use blockchain technology as a tool to redefine a framework where information can be used for transaction purposes (PricewaterhouseCoopers, 2018).

The unique features of blockchain have opened up new opportunities in various sectors such as supply chain, energy, food, retail, healthcare, education and government. The impact of change will be brought to the fore through blockchain technology in the delivery of government services and Internet performance.

> "Blockchain technology can strengthen a new type of Internet, 'Internet of value'. It can also enable our cities to function more efficiently, thus increasing productivity and economic growth"
>
> **(PricewaterhouseCoopers, 2018)**

Various departments, even though working under a single government, have no cross-communication and prefer to work in silos. They present different identities to the consumer on the front end. Blockchain intervention can provide an identity management model for this. It can enable the secure movement of controlled data from the government to the citizen. Utilizing blockchain, via smartphone applications, citizens can view their public service identity and share significant data to access this public service easily. The government's role here would be that of a verifier rather than a controller, thus recreating the relationship between government and citizen. It will also improve aspects of service delivery, ultimately leading to citizen satisfaction (Borrows & Harwich, 2017).

Governments can influence this technology to gain:

- Raised level of transparency.
- Increased working efficiency.
- Trust and accountability in government's function.
- Help in securing critical data from tampering.

Blockchain demonstrates the potential to transform various aspects of smart cities and make them more proficient by providing information technology (IT) solutions that are crystal clear, strong, tamper-proof and proficient. Various journals and papers are trying to explore and analyse this technology and its implications for the growth of smart cities when used in fields like governance, logistics and financial transactions, where citizens have low trust in government yet high expectations (Mic & Nãsulea, 2018).

8.2 GLOBAL BLOCKCHAIN ADOPTION

Many countries recognize the potential use of blockchain in the public sector and are investing and experimenting with citizen-centric services to increase the quality of life with effective service delivery. One such effective case of blockchain adoption and implementation for government services can be seen in Estonia. Estonia is called "the most developed society in the world". It has built an efficient, safe and transparent system that saves time and money. Estonia has developed a blockchain solution for a number of government services called the keyless signature interface (KSI). Estonia uses blockchain technology to verify the integrity of government registers and details such as health care records, asset records, business records, sequence records, digital court systems, etc.

e-Estonia has saved more than 1400 years of working time each year and contributes 2% GDP annually for its digitized public services. Its adoption of blockchain technology can be seen as an outline for achieving a possible global implementation of blockchain (e-estonia, 2009).

There are various countries in the league of experimenting with this technology, like China, Singapore and Dubai, to develop their smart cities. Dubai aims to be a fully blockchain-powered country using three strategic pillars of Industry Creation, Government Efficiency and International Leadership. It wants to create database records that cannot be tampered with and wants all bill payments, licence renewals, real estate proceedings and visa applications to be transacted digitally using blockchain. A cryptocurrency called "emCash" was launched for transactions on government and non-government services (Deloitte, 2018; XISCHE, 2018).

The city of Moscow started an online voting system through a blockchain platform called Active Citizen (AC), where the public can vote on matters of city development such as new playgrounds or parks, sports complexes, additional bus routes, etc. without disclosing their identity or personal data, resulting in increased confidence and transparency in public services among citizens.

The Republic of Georgia has developed a blockchain registration of land to ensure transparent patents within the National Agency of the Public Registry (NAPR) and the people, thus reducing corruption (United for Smart Sustainable Cities, 2020).

8.3 BLOCKCHAIN ADOPTION IN INDIA

Technology has infiltrated every aspect of people's lives and work across various institutions, sectors and industries, making it simpler and connected. Thus, it increases people's expectations of the ecosystem of public services. The Indian government has continuously rolled out schemes and programmes with the vision to address this aspect; for example, it initiated the Smart Cities Mission (SCM) in June 2015, intending to accelerate development in the urban sector and to improve quality of life by providing local development and technology to use as a way to build smarter citizen solutions (PricewaterhouseCoopers, 2017).

In July 2015, India launched a "Digital India initiative" with technology being the prime focus to make service delivery efficient and robust for a digitally empowered nation. At the present time, many emerging technologies can boost economic development and provide a better quality of life to citizens, but blockchain is one technological breakthrough that is considered impactful across industries (PricewaterhouseCoopers, 2018).

8.3.1 Blockchain: The India Strategy

In January 2020, the government of India's think tank NITI Aayog released a proposed discussion paper entitled "Blockchain: The India Strategy – Towards unifying Ease of Business, Easy of Living, and ease of Governance". This paper initially analyses the value of the blockchain in giving trust in government and private sector partnerships, followed by observations on blockchain use cases, potential acquisition challenges, and studies of NITI Aayog driver experience in blockchain use, indicating potential cases for the ecosystem. It specifically talks about leveraging blockchain to create a new system for land records management, the pharmaceutical drugs supply chain and educational certificates to fight fraud.

In the development of one use case, NITI Aayog partnered with Gujarat Narmada Valley Fertilizers & Chemicals Limited and is functioning in the direction of executing a proof-of-concept (PoC) application that is exploring the use of blockchain expertise for fertilizer subsidy administration. These technologies will benefit from reduced time distribution of grants, transparency, eliminating the need for more approvals and distribution of documents, and the effective marketing and distribution of fertilizers. It will enable NITI Aayog to propose policy recommendations and actions to strengthen the funding mechanism, making it transparent and rewarding (NITI Aayog, 2020).

8.3.2 Vajra Payment Platform by NPCI

Due to the lack of a regulatory framework around its implementation, India initially banned the usage of bitcoin/cryptocurrency, but in March 2020 the ban was lifted as a positive move towards the financial sector. Recently, the State Bank of India (SBI), India's major government bank, in cooperation with a few other commercial banks and companies, has launched a blockchain initiative called *BankChain*, whereby it intends to reduce fraud and make transactions more transparent in the banking sector (RP, 2018).

National Payments Corporations of India (NCPI) recently announced the launch of a blockchain-based payment platform called Vajra. Vajra is a permissioned blockchain platform, where only the registered members within the network can become part of the blockchain network.

There are three nodes in Vajra:

1. **Clearing House node:** This node administers the platform. It is controlled and maintained by NCPI.
2. **Notary node:** This node provides validation only if Aadhar biometric is used for authentication.
3. **Participant node:** These nodes are represented by the banks that can post, receive and see transactions.

8.3.2.1 Process

The Clearing House node has the authority to add new nodes on the platform. Every external party that engages with the participants of the platform will be verified by the node. Vajra deals with the application programming interface (API) interactions with key management and distinct security measures and security in data access.

8.3.2.2 Benefits

- The Vajra payment system offers the following features:
- Reduction in manual processing, as Vajra is an automated payment system.
- Fewer disputes due to the decentralized and centralized data storage service.
- The platform uses cryptography to provide highly secured payment transactions.
- All transactions are transparent.

8.3.3 Indian Initiatives in Government Services

Blockchain, being a disruptive and emerging technology in India, is at a very inceptive stage. Large-scale implementations in this field are yet to be seen, but they are slowly penetrating within the industry through pilot testing and use cases. About 50% of the states in India are experimenting in the field of blockchain, especially its adoption in the public sector. The leading use cases are seen in the land registry, farm insurances and digital certificates (Nasscom, 2019). Table 8.1 depicts a summary of all the projects/initiatives happening in India in various phases along with their blockchain category.

The state of Karnataka has implemented blockchain to store educational certificates on a semi-public platform, which gives authority to the student to allow access to their documents. Maharashtra has done a small but successful pilot on a private platform for land records, while West Bengal implemented blockchain on a public platform to issue birth certificates.

There are also states like Telangana that have developed PoCs in the field of land registration and data protection. Andhra Pradesh has completed PoCs in land records management and educational certificates, while the state of Goa is exploring the feasibility of blockchain implementation in the land registry. Various states are yet to implement blockchain but have announced its inception: Uttar Pradesh will experiment in the field of land and revenue records, while Gujarat will work in the sector of e-governance.

TABLE 8.1
Various Blockchain Initiatives in India

State	Project	Status	Category
Karnataka	Educational certificates	Implemented	Semi-private
Maharashtra	Land records	Pilot completed	Private
West Bengal	Birth certificates	Implemented	Public
Telangana	Land registration, data protection	PoC completed	Private
Andhra Pradesh	Land records, educational certificates	PoC completed	Semi-private
Goa	Land registry	Exploring feasibility	Private
Uttar Pradesh	Land and revenue records	Announced	Private
Gujarat	e-Governance	Announced	Semi-private

8.3.4 Blockchain-based Document Services

Blockchain being a nascent technology in India, its applications are slowly penetrating into the system in the form of pilot projects. Its major implementations are for maintaining land records and certificates, since they require more transparent and immutable data collection to increase trust among the stakeholders and offer an easy flow of information for efficient service to the public. Some of them are discussed in the following paragraphs.

In November 2019, Thane smart city within Maharashtra state conducted a pilot where blockchain technology was integrated into property tax documents. The pilot programme is a programme developed by the Department of Housing and Urban Development (MoHUA), "Accelerating the Growth of New India's Innovation" (AGNIi), "StartupIndia" and "InvestIndia". It is a QR code-based one-scan solution, which is expected to produce authentic encrypted documentation that will provide instant document verification when stored in the blockchain, alleviate tedious paperwork, and enable various stakeholders like bank executives, property buyers, lessees and others to authenticate the original text along with current informs, reducing fraud to the system as well as to citizens. This pilot project was recognized as the smart solution of the year in 2019 at theseventh smart cities conclave and given an award under the category of new technological innovation to improve urban public service delivery (Sanghani, 2020).

In the state of West Bengal, New Town Kolkata Green Smart City Corporation Limited (NKGSCCL) has successfully introduced blockchain technology to issue birth certificates from February 2019. The blockchain solution was introduced to generate authentic government documents with genuine data entry and all-time data availability. It has also introduced the use of blockchain technology for its online services for death registration since February 2019. By the end of 2019At the time of writing, there have been 402 birth registrations and 365 death registrations (NKGSCCL, 2019). NKGSCCL bagged the gold award in the Best Smart Health Initiative of the Year category for application and processing of the birth/date component of the electronic health record (EHR) using blockchain technology at Global Smart Cities Forum 2020 (ETGovernment, 2020).

The e-governance department of Karnataka introduced a State Scholarship Portal (SSP) to simplify the scholarship application process offered by different departments and added blockchain-based storage and security. This blockchain-based solution provides an improved experience as it makes the verification process instant and paperless. It improves transparency when documents are stored using cryptographic hashing and timestamping, creating an immutable blockchain record with a unique private key (Chandrashekhar, 2017; Government of Karnataka, 2019).

8.4 CENTRE OF EXCELLENCE IN BLOCKCHAIN TECHNOLOGY

The National Informatics Centre (NIC) and the Ministry of Electronics & Information Technology (MeitY) set up the Centre of Excellence (CoE) in Blockchain Technology in the city of Bengaluru of Karnataka state to promote blockchain and its application in January 2020.

1. The centre aims to deliver blockchain as a service (BaaS) and allow all stakeholders to benefit from learning, experience and resources.
2. It will work with global experts to guide the development and implementation of new blockchain solutions from conceptual evidence to manufacturing.
3. The centre has developed PoCs like *Blood Bank Management System*, *Public Distribution System*, *Land Record Management System*, etc., some of which are being piloted or will be shortly (CoE in blockchain technology, 2020). These PoCs are discussed in the following.

8.4.1 Blood Bank Management System

Safe access to blood and blood components is essential for a variety of critical hospital care procedures. Any contaminated blood used in the transfusion can lead to several health complications for the blood recipient.

8.4.1.1 Current Process

The end-to-end process from blood collection to transfusion is depicted in Figure 8.1.

FIGURE 8.1 Blood donation process. (From Centre of Excellence in Blockchain Technology, *Blockchain for Government*. Retrieved 7 April 2020 from https://blockchain.gov.in/Whitepaper_30jan.pdf)

8.4.1.2 Challenges

Transfusion of unsafe blood caused by human error. Insufficient blood supply management to meet demand. The unavailability of a donor's medical history can lead to a lack of transparency and traceability, as depicted in Figure 8.2.

8.4.1.3 Proposed Blockchain Solution

Introduction of a blockchain solution called a blood chain system to address the challenges and maintain an effective immutable repository. It will integrate with blood bank applications and record transactions in the chains, like donor registration, sample collection, testing, storage, blood requisition and transfusion. The system would provide privacy features, like donor information and test results, which will be encrypted and available to authorized users only. Also, it will connect key stakeholders involved in blood supply management, like donors, test centres, blood banks and hospitals.

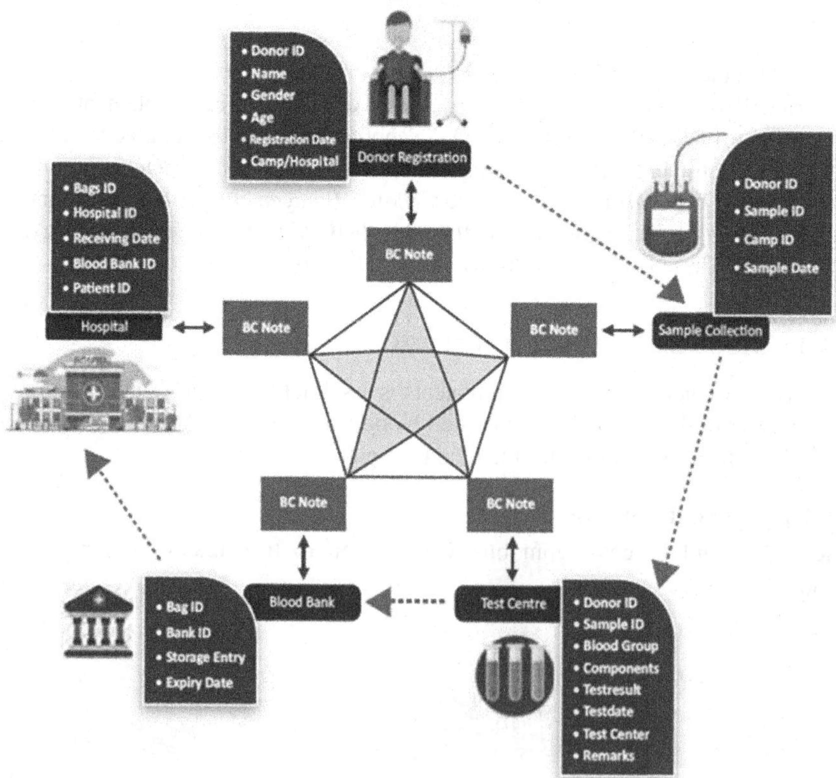

FIGURE 8.2 BloodChain system. (From Centre of Excellence in Blockchain Technology, *Blockchain for Government*. Retrieved 7 April 2020 from https://blockchain.gov.in/Whitepaper_30jan.pdf)

8.4.1.4 Benefits

Quality and availability of blood are visible to all stakeholders. Donor history can be tracked and verified. Blood is traceable from donor to recipient. Also, a lot of blood can be saved from wastage through this streamlined process, since mismanagement leads to the elimination of blood bags.

8.4.2 Public Distribution System (PDS)

PDS evolved as a system for the distribution of grains at affordable prices to the population that is below the poverty line, as displayed in Figure 8.3.

8.4.2.1 Current Process

Food grains cultivated by the farmers are obtained by the government through millers and procurement centres.

8.4.2.2 Challenges

PDS has unique challenges due to its unique system. The major challenge is the unpredictable supply of food grains due to seasonal activity, as its quantity and quality depend upon natural factors like water, weather, etc., a large number of grains are lost during logistics, and much goes missing from the warehouses. Also, there are many administration issues, like difficulty in obtaining a ration card, unavailability of stock, lack of public awareness, circulation of fake cards and no grievance redressal.

8.4.2.3 Proposed Blockchain Solution

The objective of using blockchain technology is to have a fool-proof end-to-end PDS to track food grains and record all important data throughout the supply chain, such as stakeholders as shown in Figure 8.4, price, Minimum Support Price (MSP), quantity delivered, etc. Here, timestamped receipts will be generated when the stock reaches the store. QR codes or radiofrequency ID tags will be used to track the grain bags within the supply chain to assert their quality and quantity. Smart contracts will be used to facilitate the payments to the farmers and millers as required.

8.4.2.4 Benefits

Blockchain-enabled PDS will provide quick payments to farmers and millers. Tracking stock stored at different locations and its duration along with identification of pilferages within the stock.

FIGURE 8.3 PDS supply chain. (From Centre of Excellence in Blockchain Technology, *Blockchain for Government*. Retrieved 7 April 2020 from https://blockchain.gov.in/Whitepaper_30jan.pdf)

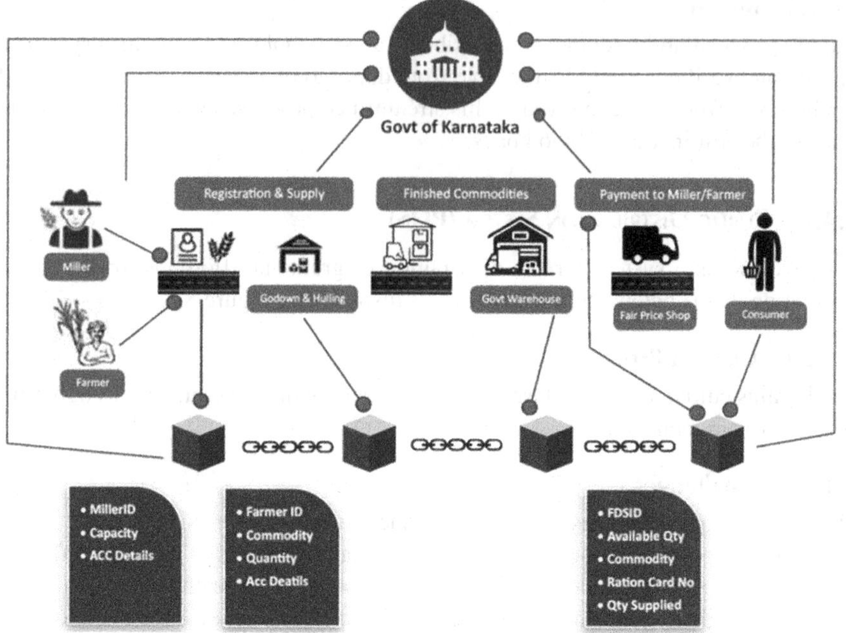

FIGURE 8.4 Blockchain-based PDS. (From Centre of Excellence in Blockchain Technology, *Blockchain for Government*. Retrieved 7 April 2020 from https://blockchain.gov.in/Whitepaper_30jan.pdf)

8.4.3 Urban Property Management System: e-Aasthi

Due to the digitization of data, many states have digitized property records and enabled transactions on a digital platform only to enable the seamless exchange of information as displayed in Figure 8.5.

8.4.3.1 Current Scenario

Workflow-based e-governance system where urban local body revenue officials enter data in the field itself during visits. It then creates digitally signed property records for more transparency. There are many contributors for maintaining the data of ownership, rights and liabilities.

8.4.3.2 Challenges

Despite the digitization of data, there are several fraudulent transactions and land disputes due to the lack of a single-source-of-truth for land ownership and the possibility of data manipulation. Also, high turn-around-time (TAT) for service delivery, which is showcased in Figure 8.6.

8.4.3.3 Proposed Blockchain Solution

e-aasthi is an e-governance system to be integrated with the blockchain to maintain a ledger by all departments. Rules to be defined for updating attributes related to the

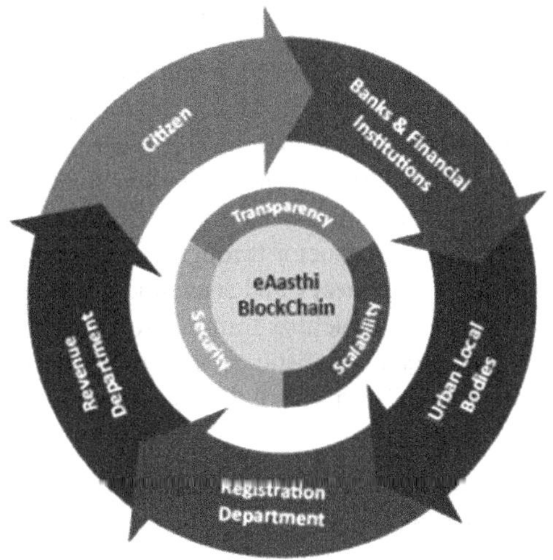

FIGURE 8.5 e-Aasthi. (From Centre of Excellence in Blockchain Technology, *Blockchain for Government*. Retrieved 7 April 2020 from https://blockchain.gov.in/Whitepaper_30jan.pdf)

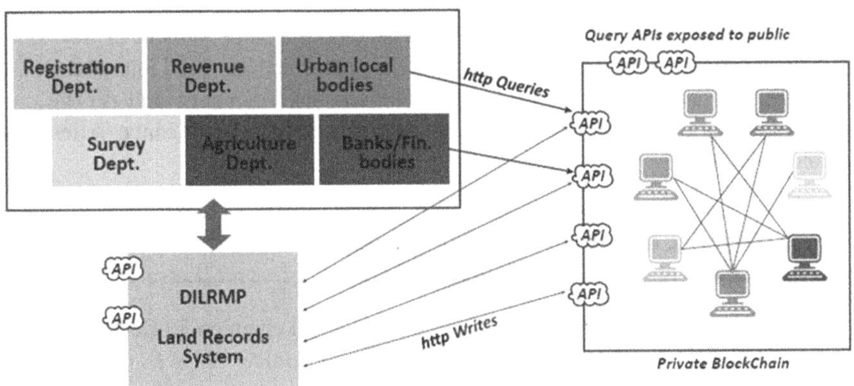

FIGURE 8.6 Blockchain-based e-Aasthi. (From Centre of Excellence in Blockchain Technology, *Blockchain for Government*. Retrieved 7 April 2020 from https://blockchain.gov.in/Whitepaper_30jan.pdf)

owner and property. Data stored in the blockchain would help provide an effective solution to citizens.

8.4.3.4 Benefits

It would create immutable property records and transactions to track the rightful owner and also help to track property ownership through history transfers. In the case of mortgaging property, banks and housing finance companies can validate the information before sanctioning.

8.4.4 Apiary – Centre of Excellence (CoE)

In July 2020, the Apiary – CoE was launched in Gurugram by Software Technology Parks of India (STPI) in partnership with MeitY, the Government of Haryana, the Government Blockchain Organization and several blue-chip companies and premium educational institutions. The centre aims to identify and boost 100 start-ups in the arena of blockchain innovation, to create innovative solutions to meet the existing and emerging challenges of the sector through mentorship and shared learning experiences from accomplished industry and academic pioneers, and to provide the highest standard of physical infrastructure, training, research & development, funding and networking. It will create a holistic ecosystem and give rise to the government of India's "Make-in-India" and "Digital India" programmes (Apiary, CoE in Blockchain Technology, 2020).

8.5 REGULATORY APPROACH

The contradiction between blockchain-based arrangements and existing legal and hierarchical structures is a significant limitation to opening the transformative potential of blockchain. Thus, there is a need for change at the policy level in order to boost the technology and bring maturity to the ecosystem of distributed ledger systems. An ideal framework for blockchain implementations policy should be; (i) learning & sharing between the states; (ii) an engaged improvement of new pilot ventures; (iii) characterizing security, protection, administration and interoperability measures; (iv) the making of blockchain fundamental segments; and (v) the creation of foundations devoted to explicit use instances of high significance (Allessie & Sobolewski, 2019).

8.5.1 National Institute for Smart Governance (NISG) National Blockchain Strategy

The NISG put forward a draft national strategy on blockchain on 30 December 2019 under the guidance of Professor S. Shivendu from the University of Florida.

The aim is to introduce a set of policy frameworks and incentives by advising stakeholders to significantly improve blockchain integration and existing environmental economic programs through:

- Architecture of legal and administrative structures.
- Creating a framework to motivate academics.
- The field of research and education promotion.
- Formulating policies that lead to faster innovations.
- The adoption and growth of blockchain in the public and private sectors.

This draft strategy recommends mainly creating laws and regulations on the basis of the functions achieved by blockchain and not on the technology itself. It suggests establishing a body to coordinate blockchain strategy across various state bodies

1. Ensure technology neutrality

3. Leadership in knowledge leads to leadership in technology

2. Ensure policy and regulatory framework at national level

4. Development of capacity in government

FIGURE 8.7 Key principles for blockchain strategy.

and also to contribute predominantly to research this incipient technology for more visionary and leading adoption (Shivendu, 2019).

The National Blockchain Strategy in reference to the following four key principles showcased in Figure 8.7 is as follows:

1. **Technology neutrality:** Development of a policy that provides a platform for competing technologies to emerge to their potential.
2. **National-level policy and regulatory framework:** Implementation of a national-level policy to remove uncertainty and encourage private sector investments. It should not restrict innovation and should consider smart contracts legal.
3. **Research and development:** The government needs to invest in universities for promising research works through proper funding mechanisms.
4. **Develop government capacity:** Important to provide an understanding of blockchain to officials and encourage and support various departments within governments to explore this technology through pilots.

This draft sheds light on promising opportunities that India can reap through the application of blockchain, such as monetization of data that is generated by the citizen and Internet of Things (IoT) devices present in the system in a secure manner and potentially contribute to the economy (Shivendu, 2019).

Table 8.2 explains the current scenario of blockchain technology in India through the Strength, Weakness, Opportunities, and Threats (SWOT) analysis.

8.5.2 State-level Policy – Tamil Nadu

On 19 September 2020, the state of Tamil Nadu became the first to release a state-level policy on blockchain. The blockchain policy envisions delivering citizen-centric services, including healthcare and a "portable digital identity".

Objectives of the policy:

a) Create a common set of guidelines to administer blockchain implementations across the Tamil Nadu Government to foster interoperability.

TABLE 8.2
SWOT Analysis of the Blockchain in India by NISG

Blockchain – Where does India Stand? A SWOT	
STRENGTHS	**WEAKNESSES**
• Large technology workforce that has been the knowledge backbone of the world can quickly reskill for leadership. • Strong identity management system across the country in the form of Unique Identification Authority of India (UIDAI) and Aadhar. • Strong IT consulting and implementation partners like National Institute of Standards and Technology (NIST), National Informatics Centre (NIC) and e-governance practices adopted.	• Lack of regulatory clarity stifling the plans of decision makers on investment programs to boost blockchain adaption. • Lack of government support for blockchain projects as a key user of the technology has delayed the takeoff. • Negligible investments in blockchain by private sector due to lack of understanding of the potential benefits and government's stand.
• Use cases across multiple domains for PoCs. • Access to global technology leaders and platforms as potential technology partners and customers offers a captive opportunity. • Education. • Ability to adopt integrated strategies across multiple disruptive technology domains.	• Very few production-level applications in country. • Investment climate not conducive to innovation as India's venture capital is mostly focused on growth-oriented projects and not for incubating. • Curriculum. • Very poor awareness among decision maker community in public and private sector.
OPPORTUNITIES	**THREATS**
• Opportunity to be the blockchain development backbone of the world by reskilling the developer population in advance. • Potentially largest pool of IoT devices generating monetizable data, as India sports one of the highest citizen bases and telecom penetrations.	• Large pool of IoT devices susceptible to cyberattacks can derail the automation programmes and Industry 4.0 plans. • Differences between different entities that are expected to collaborate may create roadblocks and deadlocks.
• Large opportunity for data monetization by creating a market place for anonymized data through blockchain can lead to unlocking of the value both in the country and abroad. • Increasing transparency in banking system to eliminate Non Performing Asset (NPA) burden. Transparent processes for procurement and loan process management can help with humongous savings.	• The high potential of data-triggered prosperity can lead to excessive attention from cyber attackers. Poor compliance on cybersecurity best practices can spur a collapse of connectivity. • The transparency-threatened lobbyist who has been used to exploiting inefficiencies in weak systems can apply roadblocks to blockchain programmes.
• Improving transparency in benefit programmes can enable the schemes to maximize positive impact on citizens. • Crashing expenses of elections and offering instantaneous election results.	• Improper connectivity of distant places and underdeveloped areas can restrict the benefits of technology depending on Internet connectivity. • Lack of digitization and legacy backlogs can create inertia to shift to advanced technologies.

Source: Shivendu, S., *National Strategy on Blockchain*, 2019. Retrieved 20 January 2020 from www.nisg.org/blockchain.

b) Build a developed and self-sustaining blockchain community.
c) Build a regulatory sandbox for building and employing blockchain applications.
d) Create an oversight mechanism for a successful rollout, implementation and adoption of blockchain (Government of Tamil Nadu, 2020).

8.5.3 Draft State Policy – Telangana

The Government of Telangana has conceptualized India's first *Blockchain District*, which will be an actual region within Hyderabad that aims to create the world's best ecosystem for blockchain technology.

Vision: To make Hyderabad one of the top ten Blockchain Cities of the world

The draft policy, which sets the strategic direction, is based around four main pillars.

1. Developing Talent Pool
2. Supporting Infrastructure
3. Promoting Research and Innovation
4. Enabling Collaboration and Building Community

It will also provide incentives to companies developing blockchain products and delivering blockchain services, as shown in Figure 8.8 (Government of Telangana, 2019).

8.6 CONCLUSION

New technology is welcomed in the ever-evolving world, as it promises to improve the existing system and bring efficiency to service delivery, giving satisfaction to citizens and stakeholders. Blockchain technology in India is at a very inceptive stage; although large-scale implementations are yet to be seen, they are slowly penetrating within the industry through pilot testing and use cases. About 50% of the states in India are experimenting in the field of blockchain, especially its adoption in the public sector, as it shows an inviting path towards more responsive services, better operational efficiency and robust data security for its citizens and stakeholders. But, adoption of new technology like blockchain requires a completely new roadmap that

FIGURE 8.8 Four main pillars of blockchain strategy.

is agile in its work flow, especially in the government sector, where it needs to be validated, regulated and adopted. The success factors of such an initiative include the collaboration of leadership, funding, government and citizens. Coordination of stakeholders like urban local bodies, the state and the central body for data sharing will be a major requirement to create a reliable database for efficient functioning through interoperability. But, the adoption of new technology like blockchain won't be without any hurdles, especially in the public sector, where it is necessary to be certified, regulated and accepted. In order to harness the power of the new blockchain and reach the level of greater implementation, a strategic approach that includes distribution policy, management, environmental development, talent nurturing, educational focus and awareness building is required to drive a mind-set for change of trust with government authorities and citizens. In conclusion, blockchain presents a promising opportunity for India, which will completely change the way government organizations work (PricewaterhouseCoopers, 2018).

REFERENCES

Allessie, D., & Sobolewski, M. (2019). *Blockchain for digital government.* Joint Research Centre. European Commission. doi:10.2760/93808

Apiary, CoE in Blockchain Technology. (2020, July 17). *Apiary, a centre of excellence in blockchain technology.* Retrieved November 22, 2020, from https://apiary.stpi.in/index.html

Borrows, M., & Harwich, E. (2017, November). *The future of public service identity:Blockchain.* Retrieved August 28, 2019, from https://reform.uk/sites/default/files/2018-10/Blockchain%20report_WEB%20%281%29.pdf

Centre of Excellence in Blockchain Technology. (2020). *Blockchain for Government.* Retrieved April 7, 2020, from https://blockchain.gov.in/Whitepaper_30jan.pdf

Chandrashekhar. (2017). *National e-Governance Division.* Retrieved February 10, 2020, from https://negd.gov.in/sites/default/files/e-AttestationusingBlockchain-Karnataka.pdf

Deloitte. (2018, January). *Blockchain in Public Sector.* Retrieved October 27, 2019, from https://www2.deloitte.com/in/en/pages/public-sector/articles/blockchain-in-public-sector.html

e-estonia. (2009). *e-estonia briefing centre.* Retrieved January 27, 2020, from https://e-estonia.com/

ETGovernment. (2020, November 28). Global smart cities forum 2020. Retrieved December 24, 2020, from ETGovernment website: https://img.etb2bimg.com/files/cp/upload-1608723395-winners.pdf

Government of Karnataka. (2019, November 11). *State scholarship portal and e-attestation.* Retrieved January 24, 2020, from https://ssp.postmatric.karnataka.gov.in/docs/new/SSPPOSTMATRICPPT.pdf

Government of Tamil Nadu. (2020, September 19). *Tamil Nadu blockchain policy 2020.* Retrieved December 24, 2020, from tnega.tn.gov.in: https://tnega.tn.gov.in/assets/pdf/block_chain_2020_final.pdf

Government of Telangana. (2019). *Draft blockchain policy.* ITE & C Department. Retrieved January 13, 2020, from https://it.telangana.gov.in/wp-content/uploads/2019/05/Telangana-Blockchain-Policy-Draft-May-2019.pdf

Mic, S.-M., & Năsulea, C. (2018, May). Using blockchain as a platform for smart cities. *Journal of E: Technology, 9*(2), 37–43. Retrieved from https://www.academia.edu/37326061/Using_Blockchain_as_a_Platform_for_Smart_Cities

Nasscom. (2019, March 13). *Avasant-Nasscom India blockchain report 2019.* Retrieved October 27, 2019, from Nasscom Web site: https://www.nasscom.in/knowledge-center/publications/nasscom-avasant-india-blockchain-report-2019

NITI Aayog. (2020). *Blockchain: The India strategy.* Retrieved February 5, 2020, from https://niti.gov.in/sites/default/files/2020-01/Blockchain_The_India_Strategy_Part_I.pdf

NKGSCCL. (2019, September). *Case study on birth registration on blockchain technology in Newtown.* Retrieved February 26, 2020, from New Town Kolkata Green Smart City Corporation Limited: https://www.newtowngreencity.in/wp-content/uploads/2019/09/Blockchain_29.07.2019_Brochure.pdf

PricewaterhouseCoopers. (2017). *Digital India: Targeting inclusive growth.* Retrieved March 15, 2020, from https://www.pwc.in/assets/pdfs/publications/2017/digital-india-targeting-inclusive-growth.pdf

PricewaterhouseCoopers. (2018, November 21). *Automating trust in citizen services.* Retrieved April 25, 2020, from https://www.pwc.in/assets/pdfs/publications/2018/automating-trust-in-citizen-services.pdf

RP, S. (2018, July 25). Bank-Chain: India's first Blockchain exploration consortium launched for banks. *Blockchain.* Retrieved January 24, 2020, from https://www.expresscomputer.in/news/bank-chain-indias-first-blockchain-exploration-consortium-launched-for-banks/20453/

Sanghani, P. (2020). *The Economic Times.* Retrieved February 15, 2020, from https://economictimes.indiatimes.com/news/politics-and-nation/thane-municipal-corporation-turns-to-blockchain-for-property-tax-assessments/articleshow/73183406.cms

Shivendu, S. (2019, December). *National strategy on Blockchain.* Retrieved January 20, 2020, from https://www.nisg.org/blockchain

United for Smart Sustainable Cities. (2020, November). *Blockchain for smart sustainable cities.* Retrieved December 1, 2020, from ITU-T: https://www.itu.int/pub/T-TUT-SMARTCITY-2020-54

XISCHE. (2018). *State of play blockchain.* Xische & Co. Dubai: Creative Commons. Retrieved 7 25, 2019, from http://insights.xische.com/XischeReports_StateofPlay_Blockchain.pdf

9 Role of Blockchain in the Music Industry

V. Srividya and B.K. Tripathy

CONTENTS

9.1 INTRODUCTION

In today's world, the amount of data created and shared over the network has grown from terabytes to zettabytes(ZB). Every minute, around 480,000 tweets are constructed, 4.7 million videos are viewed, 400 new users are created on Facebook, 60,000 images are uploaded on Instagram, 200 million emails are sent and 4.2 million search queries are processed (NodeGraph, 2020). The amount of data available on the digital universe has been reported to be around 2.7 ZB as of 2017. The data being addressed here is created, captured or replicated but is not being stored.

DOI: 10.1201/9781003138082-9

The traditional way of storing data was in a database, either in a structured format, referred to as the Relational Database System (RDBMS), or in an unstructured format. Structured data would include information that could be stored in the form of rows and columns, like address book details, booking details and so on. Unstructured data consists of images, audio, video, Java Script Object Notation (JSON) files and many more.

The key principle of a database is that it is centralized in nature and is controlled by the administrator. The architecture used is the client–server architecture. There is a chance of the data being tampered with by malicious users. Only the administrator gives access to the users for any specific tables of the database.

There are many security concerns that arise when databases are used that could compromise the integrity of data (Preethi et al., 2020). Wrong access privileges or unchanged privileges could allow unauthorized users to modify or copy data from the database. Some malware that affects the computers of the users might get access to the database via the connected networks. Databases that have not been maintained properly are vulnerable to exploitation. Malware, or some code that would grant remote access to the database, could be both SQL based and NoSQL based.

In 2008, Satoshi Nakamoto developed a technology called "blockchain" that would serve as a public transaction ledger for the cryptocurrency "bitcoin".

9.2 BLOCKCHAIN

Blockchain is a timestamped series of immutable records of information that are linked to one another in the form of a chain (Rosic, 2020). The fundamental feature of blockchain is the fact that it is not managed by any individual. The links between blocks are created using cryptographic information such as hash functions. This feature helps in maintaining the integrity of the data stored in the blocks.

Integrity can be classified as a non-functional feature of any software system (Drescher, 2017). It consists of three major components:

- Data Integrity: Data that is used and maintained by the system must be complete and correct and must not have any inconsistencies.
- Behavioural Integrity: The system must behave as intended and also must be free from any logical errors.
- Security: Restricted access of data and its functionality to only authorized users must be ensured.

Figures 9.1 and 9.2 show a single block of the blockchain. Figure 9.1 shows the structure of the block, and Figure 9.2 shows the block header.

As seen in Table 9.1, there are significant criteria for the assessment of chemical fuels.

The block number associated with each block is a unique identifier that can be used to distinguish blocks from each other. The block size field contains the total size of the block, including the size of the block header and the cumulative size of the

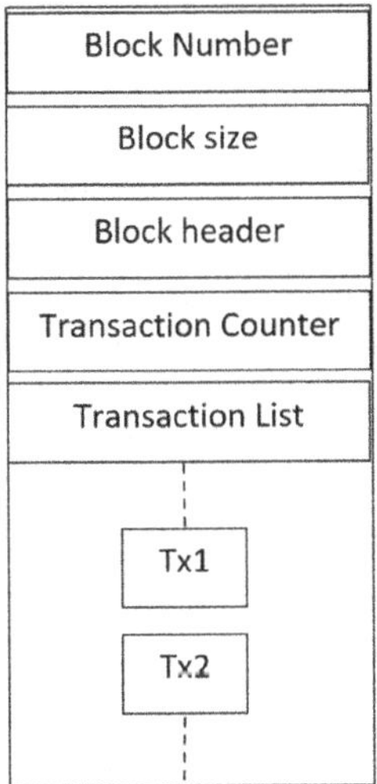

FIGURE 9.1 Structure of a block.

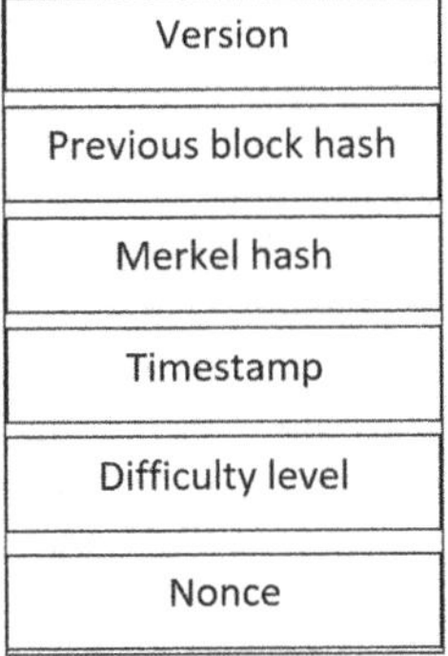

FIGURE 9.2 Structure of a block header.

transactions of that block. The total size of the block is around 1 MB. The transaction counter contains the number of transactions included in the block. The size of this field varies between 1 and 9 bytes. The transaction list stores the digital fingerprint of all the transactions that the block comprises.

The contents of the block header include the following:

- Version: It is a 4-byte field, which specifies the version number used in the block. The version value specified in this field has a corresponding set of validation rules that the block will have to follow. In a blockchain, all the blocks must have the same version value in their header. If the block contains different version details, a "hard fork" occurs. Table 9.1

TABLE 9.1
Version Types

Version Type	Description
Version 1	This was the first version that was used when blockchain came into existence. It is known as the genesis block and was created in January 2009.
Version 2	Corresponds to bitcoin core 0.7.0, used the User Activated Soft Fork (UASF) and was created in September 2012. It has been described in BIP34.
Version 3	Corresponds to bitcoin core 0.10.0, used UASF and was created in February 2015. It has been described in BIP66.
Version 4	Corresponds to bitcoin core 0.11.2, used UASF and was created in November 2015. It has been described in BIP65.

shows the different version types. The version numbers can be any of the following:

- Previous block's hash: It is a 32-byte field. The value in this field, as the name suggests, is the hash value of the previous block's header (Tabora, 2019). This acts as a pointer to the previous block in the blockchain. One of the interesting features of this field is that it cannot be changed or manipulated without changing the details of the previous blocks. If an intruder wishes to change the previous block's header in a large blockchain, he or she must change every previous block prior to the one being modified. This is known as the "Avalanche Effect". The cost involved in altering so many blocks is very high, as it would require massive amounts of computing power to calculate the hash rate. This is the key feature that makes the blockchain secure and resistant to any tampering. SHA-256 is the hashing algorithm used to create a hash of the previous block.
- Merkle hash or Merkle root: The size of the field is 32 bytes. The content of this field is the hash value of the blockchain's Merkle root. The hashing algorithm utilized is the SHA-256. The Merkle root of a block is calculated using the hash values corresponding to each transaction that has been included in that block. It consists of all the transaction IDs of that block. The placement or order of the transaction IDs must satisfy the consensus rules.
- Timestamp: This is a field that can accommodate data of 4 bytes. The time specified in this field is a measure of the UNIX epoch time. The value of the timestamp can be defined as the elapsed number of seconds since January 1970. The timer starts when the miner begins the hashing process of the header. A timestamp is said to be valid if it is found to be greater than the median time of the preceding 11 blocks.
- Difficulty level or Difficulty target: This field is also referred to as "bits" and has a size of 4 bytes. The encoded bits act as a threshold value, which must be equal to or greater than the hash value of the block's header. The

functionality of this field is that it governs the level of difficulty of the hash value such that the network's total hash rate is satisfied.

- Nonce: This field's size is 4 bytes. The value in this field can be described as an arbitrary number that can be adjusted by miners such that the block's hash is less than or equal to the current target of the network. The block validator is the first miner to discover the nonce value. The validator is rewarded for his find.

9.3 BITCOIN

Generally, when "blockchain" is addressed, the term "bitcoin" is also a part of that sentence. In 2008, the blockchain technology was developed to support and maintain the transactions of the "bitcoin" cryptocurrency. It uses a peer-to-peer technology to facilitate the transfer of funds through instant payments.

9.4 FUNDAMENTAL TECHNOLOGIES OF BLOCKCHAIN

The fundamental technologies of blockchain as shown in Figure 9.3 include a publicly distributed ledger, encryption using cryptographic keys, a peer-to-peer network sharing a common ledger and finally, a way to compute and store transactions and records of the network.

A public distributed ledger can be defined as a digital system that allows users and systems to store transactions. It originated from the peer-to-peer network. The features of the ledger include immutability, decentralized nature, distribution among users who have full or partial access to it, and finally, data can be added to the ledger only as a new entry; in other words, any modification to the existing data or adding new data is done in append mode only. There are three types of distributed ledgers:

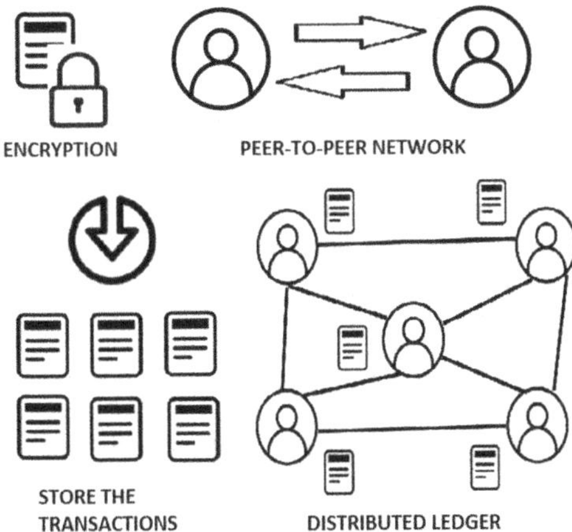

FIGURE 9.3 Fundamental technologies of blockchain.

Permissioned, permission-less and hybrid. The first one requires that the user request access to the network's functionality or features; the second type does not require any permissions, and any user can get access to the network and its features. The hybrid type is a combination of the first two types.

The process of encryption using cryptographic keys is done to create digital signatures for each blockchain. These digital signatures help in providing integrity to the entire process involved in transferring data over the network through blockchains. The type of cryptography used in blockchain is asymmetric cryptography that involves a private key and a public key. The private key is used to encrypt the message, and the public key is shared with the public; the users need to use their private key along with the public key to decrypt the message.

A peer-to-peer network is a network that does not involve any intermediate servers while sharing data between two or more users (peers). Each peer has equal privileges to the other. This forms the backbone of the blockchain technology.

9.5 OPERATIONS INVOLVED IN THE WORKING OF BLOCKCHAIN

The operations that are involved in the working of blockchain are hash functions, proof-of-work and mining. The hash functions are used to encrypt the data in the blocks before they are sent through the network in order to prevent any unauthorized access to the blocks. The hashing algorithm used is SHA-256. This algorithm creates a 256-bit signature for a text. Every user has two keys, a private key and a public key. The public key helps to uniquely identify the user, and the private key gives the user access to their account. With respect to the block, the previous block's header is hashed, the transactions in the block are hashed, the nonce is hashed and the block ID is hashed. The block ID is generated using the previous block header's hash, the transaction detail's hash and the nonce's hash values.

Data that consumes a lot of time or money but can be easily verified by other peers such that it satisfies the requirements is termed *proof-of-concept.* In general, miners have to find the nonce value for a block by solving some puzzle that involves mathematics. Once the nonce value is solved, the miner shares it among the network. Once the other miners verify it, the miner who cracked the nonce will be rewarded.

Miners keep solving mathematical puzzles to decode the nonce value of the block. Whichever miner is the first to get the nonce value is rewarded. This process is termed *mining.* The winning miner is rewarded with bitcoins after the nonce he found is verified using the proof-of-concept. The fee associated with addition of blocks to the blockchain is 12.5 bitcoins.

9.6 PROPERTIES OF BLOCKCHAIN

Table 9.2 describes some of the key properties of blockchain.

9.7 APPLICATIONS OF BLOCKCHAIN TECHNOLOGIES

Table 9.3 shows the different areas of applications of blockchain technology (Daley "Block Chain", 2020; Daley "Health Care", 2020).

TABLE 9.2
Properties of Blockchain

Property	Description
Decentralized technology	The network does not have any single agent who controls the operation; instead, a group of users maintain the network.
Enhanced security	Since the contents of the blocks of a blockchain are encrypted, no malicious user can easily modify the contents of the block.
Distributed ledgers	The ledger is available to all the users of the network, and the computational power is distributed to all the computers, resulting in better yields.
Immutability	Every user has a copy of the ledger with them. Whenever a user wishes to add data to the ledger, it must be validated by the other users in the network before it can be added to the ledger.
Faster settlement	The transfer time involved in blockchain is lower compared with traditional means such as banking.

TABLE 9.3
Applications of Blockchain Technology

Application	Description	Companies that have implemented blockchain
Secure sharing of medical data	Works on the concepts of smart contracts that control the sharing of private medical data between patients and doctors	BURSTIQ, FACTOM, MEDICALCHAIN
Real estate	Registering real estate properties using the decentralized mechanism of the blockchain	PROPY
Finance industry	Transfer of cryptocurrency in a safe and efficient manner	CHAIN, CHAINANALYSIS, CIRCLE

9.8 INTRODUCTION TO MUSIC INDUSTRY

Music can be used to describe a person's traits, as in general, music is about personality. Music has a varied variety of genres, such as pop, folk and culture, hip hop, devotional, romantic, party, dance numbers, instrumental, RnB, rock, jazz, soul and so on. Each user has his or her favourite genre of music, which they tend to listen to in a loop, or depending on their mood at that instance, they may choose certain genres. Based on these choices, one can identify the state of mind the user is in or what kind of person he or she is.

Music is something that is listened to all the time, be it while travelling or working out or cooking, to name a few situations. In short, music plays an eminent role in one's life. The industry of music is a global industry, having different forms across the globe. The next section describes the supply chain of the industry and the roles of all the stakeholders in the chain.

9.9 THE MUSIC INDUSTRY SUPPLY CHAIN

Players involved in the music industry are the artists, publishers, retailers, labels or record companies, streaming digital service providers, and performance rights organizations (PROs), as shown in Figure 9.4 (De León and Gupta, 2017).

The authors and the performers are the starting point of the supply chain of the music industry. These two parties are responsible for creating and performing the music. Each of these stakeholders is given different treatment when it comes to the copyright laws. In the music industry value chain, the authors and the performers are the ultimate holders of the rights. The income received by them based on their licence agreements is not very high, as their negotiating skills are highly variable.

Labels are the next in the supply chain industry and are responsible for managing the recording, finance, promoting the music in the market and so on. Distributors are responsible for shipping music to retailers and digital streaming partners like Amazon Music, Spotify and Gaana, to name only a few. The retailers are the last in the supply chain, who keep the CDs and USBs on their shelves. The money from the user goes to the retailer, and then the amount is distributed to all in the supply chain based on their agreements.

The intermediaries involved in the supply chain are the labels (T-series, Raaga, Tips, Sa Re Ga Ma and so on) and the PROs. The task of these two entities is to provide the recording instruments, operational support and marketing, managing and monitoring the intellectual property (IP) and monetizing the IP by managing the royalties involved.

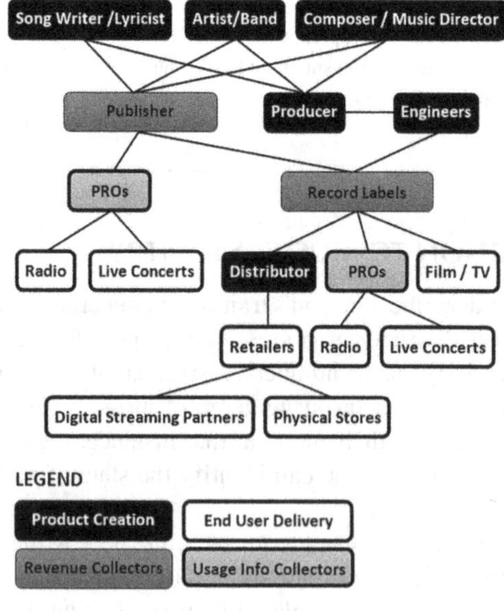

FIGURE 9.4 Music industry supply chain.

9.10 DIGITAL STREAMING OF MUSIC AND ITS IMPACT ON THE MUSIC INDUSTRY

With the emergence of the Internet era, the supply chain of the music industry has been rearranged. This new plan has accelerated the access to fresh songs reaching the market with the help of streaming services such as Spotify, Saavan, Gaana, Wynk and so on (Arcos, 2018). The focus has drifted from owning the music to accessing the music, which in turn has drastically affected the monetization procedure involved in the industry.

Before moving on to the losses that have been encountered due to the elimination of some creators and consumers from the supply chain, let us understand the areas in the music industry and also in what ways people can make an earning from these areas.

Considering a single or an album being created, the following are the types of rights that are available with the album (Horus Music, 2020).

- Song rights: Every song before reaching the target market must be written, composed, published and also performed. Each of these tasks has rights associated with it. All these rights fall under the parent category of song rights. These are generally owned by the artist or the band and can be split among the members of the band if needed.
- Recording rights: Once the song is written and composed, it must be recorded before moving to the next stages of publishing and performing. These rights include music video rights and mastering rights. The owners of recording rights are the recording labels who pay for the recording of the songs.
- Artistic rights: Some songs in the album are depicted in beautiful locations with mesmerizing properties to enhance the creativity in the song. The photographers or the artists who contribute to the making of these music videos are the owners of the artistic rights.

9.11 MAKING MONEY FROM THE ALBUM

Suppose a person has all the rights of the song except the one reserved for the artist. He or she can make money from the album by:

- Re-recording the song. A song is divided into two segments, the composition and the sound recording. When re-recording or re-creation of an existing song is done, either of the two aspects can be modified to create a new version.
- Distributing the song to a wider audience through online streaming partners. Some of the distributors available in the market are Landr, TuneCore and Amuse. Allowing the songs to be played in public, either in public concerts or in public events such as corporate events or functions.
- Synchronizing: The recorded songs are used in television series as background songs or in films or advertisements or any form of visual media. These songs can also be broadcast via the radio.

- Selling band merchandise online: Many popular and upcoming bands or musicians sell merchandise such as sweatshirts or t-shirts with the band logo or a line from a famous song, caps, magnets, backpacks, notebooks, laptop covers, badges and so on.
- Selling beats or sub parts of a song: Some beats or sample notes from recorded songs are sold to other artists, such as DJs and sound engineers, who mix these beats to get a remixed version from a bunch of other songs.

9.12 IMPACT OF THE INTERNET ON MONETIZATION IN THE INDIAN MUSIC INDUSTRY

There has been a major impact because of Internet usage and high usage of the streaming services worldwide. This chapter focuses on the impact in the Indian market.

The structure of the Indian music industry can be divided into three main core business entities: The live industry, the recording industry and the publishing industry. In the next few lines, revenue collected in each of these three business units is explained. The live industry is wide-ranging and disjointed, and hence, precise details are difficult to get. Based on the data from the Indian Music Convention, PwC data and the discussions from industry, the total revenue in live business is \$280 million. Out of the \$280 million, \$84 million is accounted for by ticket sales, and the remaining \$196 million is from the collective revenues obtained from brand sponsorships, private events, merchandise sale and so on. The recording industry is the one that deals with the recording labels, like T-series, Sony-music, Lahari Music, Tips, Times Music and Venus, to name a few. Reports from the International Federation of the Phonographic Industry (IFPI) say that the revenue generated by the recording industry in 2018 was \$153.1 million, of which \$105.64 accounts for revenue obtained from streaming out of the total recording revenue. Around 80% of the music recorded is for films. The publishing industry is still in its initial stages; a proper publishing pipeline is under development. As of 2018, the songwriters' and composers' royalties make up less than 1% of the total industry's revenue.

The total estimate of the revenues from all the three business units adds up to \$443 million. Figure 9.5 below shows the distribution of the total revenue collected from the music industry.

The two historical players, Gaana and Saavan (JioSaavan – 2018 merger), had their highest number of streaming users around 2015–2016. Until 2015, the digital revenues had been stagnant for the past 5 years. From 2015 to 2018, there was a growth of 210% in the digital revenues owing to the sudden surge in the number of people using the Internet, thus making the streaming services an established player in the ecosystem. According to one report, as of December 2018, the number of users of streaming services in India was 150 million. Figure 9.6 shows the trend of the digital revenue from 2013 to 2018 in millions of US dollars.

In recent years, many music streaming services have become popular in India, some of them being Gaana, Amazon Music, Wynk, YouTube Music, Google Play

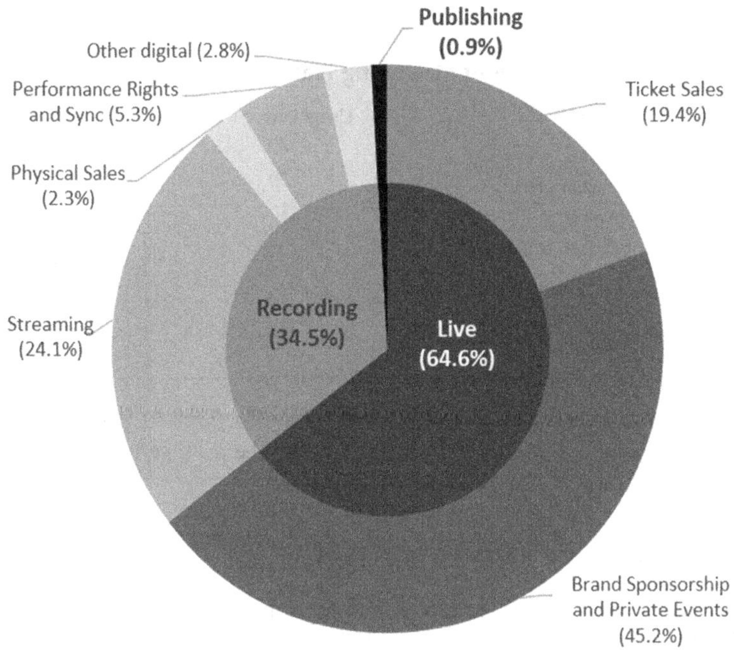

FIGURE 9.5 Distribution of total revenue from music industry.

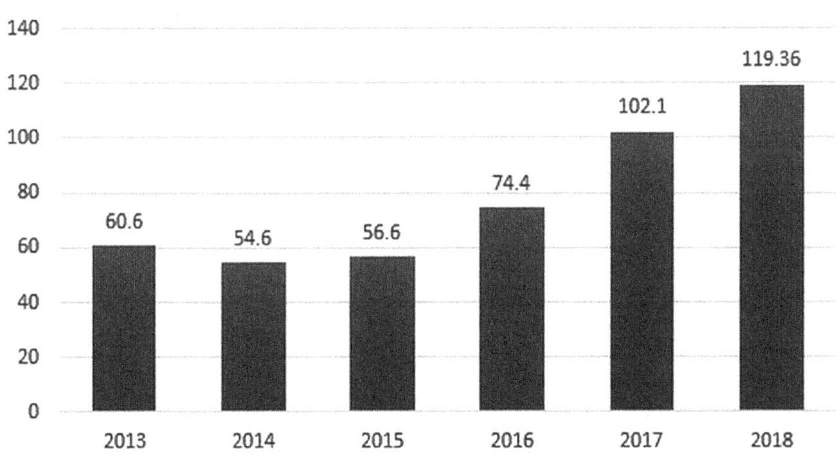

FIGURE 9.6 Digital revenue year wise.

TABLE 9.4
Usage Share of Streaming Platforms as of 2019

Platform	Usage Share (%)
Gaana	20
Amazon Music	20
Wynk	15
YouTube Music	10
Google Play store	15
Hungama	10
Others	10

store and Hungama (Ucaya, 2020; Statista, 2020). The average usage share of each of these platforms as of 2019 is shown in Table 9.4.

9.13 BLOCKCHAIN AND MUSIC INDUSTRY

Although users or customers nowadays can get songs or music streamed or downloaded with a click of a button, the people involved in its creation do not get their dues that easily. The payment to these stakeholders is not transparent either.

Stakeholders such as the songwriters, producers or musicians are the ones who get their payments last, even though they are the people who start building the product. Sometimes, these participants are not aware of how their royalties are calculated (Moreira, 2020; Kim and Kim, 2020).

This problem can be solved using blockchain technology. A common protocol will be followed, which all the members of the music industry supply chain must adhere to. Every individual will be informed about their earnings. A peer-to-peer network consisting of not only the creators of the work but also those involved in its path to reach the audience will be aware of all the details available to them in the form of a distributed ledger of metadata.

9.14 COMPANIES THAT USED BLOCKCHAIN FOR THE MUSIC INDUSTRY

In this section, we present some companies that are associated with the music industry and use the blockchain mechanism as a medium for maintenance to distribute royalties to various stakeholders.

9.14.1 MEDIACHAIN

A New York-based peer-to-peer company, which provides a blockchain database that can be used to share data among diverse applications. The work done by this company is not limited to organizing open-source information with unique identifiers but also involves working with the artists by helping them to get their fair pay, as they are the source of the music or song that is downloaded or streamed. The concept of smart contracts is used to ensure this fair payment method. There will be no third party or contingencies involved in the payment process. The smart contract will bind the musicians with only those who are directly involved in the value chain. In 2017, Spotify acquired Mediachain to help resolve the royalty payments and rights issues within the music industry.

Smart contracts can be described as computer programs that run in the blockchain network (Wang et al., 2019). The task of these programs is to trigger or apply constraints or business logic to ensure the smooth functioning of the overall task. In this case, the overall task is the process of royalty payments. The working of this platform is shown in Figure 9.7 (Mediachain, 2020).

The data from publisher to subscriber follows the procedure described in the next paragraph. The publisher will have his or her data in their system, and metadata of that data, along with a statement envelope, is the information that is passed on to the next stage. The statement envelope consists of a "To" address. It comprises two parts: The namespace and the metadata fingerprint. The namespace can be defined as the topic whose related information can be looked up from the directory. Every unit of the user-generated metadata is associated with a unique identification value (fingerprint). This unique fingerprint is doubled, making the data more secure in the process of finding or retrieving it from the database. The "From" part of the metadata consists of a signature part. Mediachain's task is to integrate identity services like Blockstack or Keybase to link the cryptographic signatures with familiar names.

9.14.2 UJO

This company, located in New York, provides a decentralized database that stores the ownership rights of music and is also involved in automating the royalty payments. Artists of the original creations, such as songs or albums, can upload these works, publish their work or control the options related to licensing and also can manage the distribution of their work through the platform based on blockchain

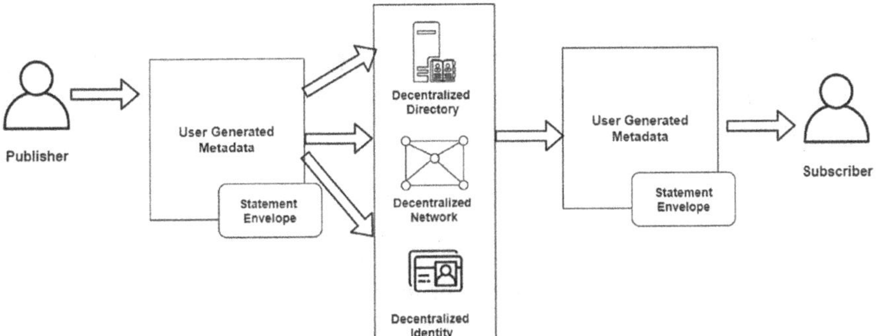

FIGURE 9.7 Mediachain workflow.

technology (Shrestha et al., 2020). The Ethereum platform is used to help remove any confusion related to ownership of the music or any payment-related issues. The concepts of smart contracts and cryptocurrencies have been used by this business.

Initially, when blockchain technology was invented, the cryptocurrency that was a part of the transactions was bitcoin. The hashing algorithm used was SHA-256. But bitcoin did have limitations, such as scalability and computation power, to name a few.

The Ethereum technology overcomes the limitations of bitcoin, and the cryptocurrency associated with it is called Ether. The main feature of the Ethereum network is the Greedy Heaviest Observed Subtree (GHOST) protocol. One main advantage of Ethereum blockchain over bitcoin blockchain is that the Ethereum block contains the most recent state of the block along with the block header, nonce, difficulty target and transaction list (Vujičić et al., 2018). The centralization problems that would occur due to stale blocks are overcome in this technology.

9.14.3 CHOON

This is a blockchain-based platform for music streaming with digital mode of payment, located in the UK. The platform provided uses Ethereum technology as the basis for smart contracts between the artists and contributors of every song. The creators of the song receive their payment within a time based on the number of streams that have been recorded, whereas with the traditional procedure, they would have had to wait for a year and might not have received the correct royalty either.

In addition to rights or royalty payments, the blockchain also provides crowdfunding to new forthcoming artists, and rewards are given to listeners when they create their own personalized playlist.

9.14.4 Open Music Initiative (OMI)

This is a non-profit business that includes around 200 members residing in Boston, Massachusetts. The aim is to have an open-source protocol that can be followed in the music industry (Sitonio and Nuccarelli, 2018). The use of blockchain technology helps to uniquely identify the music rights owners and creators, enabling them to receive their deserved royalties. The platform provided by OMI is an application programming interface (API) that helps the stakeholders of the organization in developing their own systems.

9.14.5 Musicoin

This is a platform located in Hong Kong that not only streams music but also supports the creation, use and distribution of the work in a shared economy (Sitonio and Nuccarelli, 2018). The global currency it uses, called MUSIC, provides support with respect to any trade that involves music or purchases that are related to music. Middlemen such as third parties are completely removed, ensuring that 100%

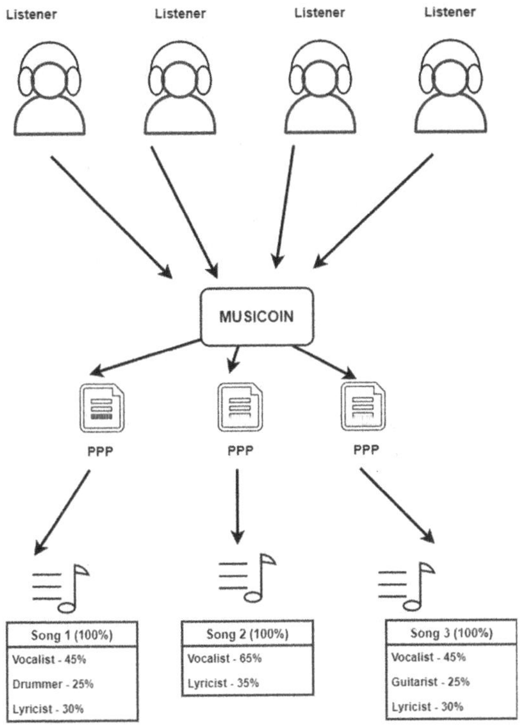

FIGURE 9.8 Musicoin working procedures.

revenue from streaming is given to the artists and contributors. The model used by this platform is known as a universal basic income (UBI). UBI makes sure each contributor is rewarded as per their contribution to the music. Payments are directed using the smart contracts concept.

Every time a song is streamed, based on the pay per play concept, a specific amount is provided to the contributors of the song. Suppose a song has three contributors; their payment is distributed based on their percentage of contribution to creating the song, as shown in Figure 9.8.

The users of the Musicoin are allowed to tip the musicians with the Musicoin currency. This can be done only when the user buys a bitcoin and gets it converted to Musicoin at the cryptocurrency exchange.

9.15 PROPOSED SYSTEM FOR USING BLOCKCHAIN IN THE INDIAN MUSIC INDUSTRY

As discussed, a lot of income in the Indian music industry is obtained from streaming or downloading music. Streamlining the procedure using blockchain technology would definitely be an added advantage not only in terms of increased revenue but also for the stakeholders involved in the industry. Every contributor will be aware of what happens to their work and how their royalty is calculated. The proposed approach to include blockchain is shown in Figure 9.9.

Consider that a listener is associated with the application (proposed platform). Every song the user listens to has its set of creators or contributors. In the scenario given in Figure 9.9, four stakeholders are directly involved with the song under consideration: The artists such as singers or actors (video song), the song writer or the lyricist, the music composer and the producer of the song. Using the Ethereum blockchain technology as a backbone for streamlining the industry, every song that is uploaded by the artist or the creator is assigned a unique identification number. This unique ID is used to associate any transaction on that song. Each song is linked to a smart contract between the creators and the consumer without any external parties. A digital wallet is also attached to the song. Any payments involving the song are made through the digital wallet using a personal cryptocurrency called "Mol", an Indian term for cost or price. The basic principle used for payment can be Pay Per Play (PPP). Each song's "Mol" amount is decided before the artist uploads his or her work. The fixed "Mol" is distributed to the contributors based on their input while creating the song.

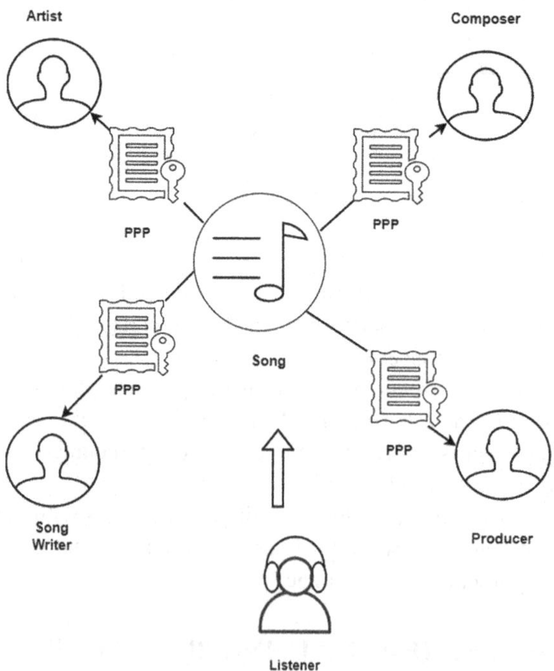

FIGURE 9.9 Proposed system working.

9.16 CONCLUSION

Music has always been a very important part of everyone's life. The modes of listening to it have changed from physical devices such as cassettes or CDs or USBs to online streaming. The number of users streaming or downloading music has been booming since 2015. The revenue of the music industry has also increased, but the people involved in the value chain do not receive their dues on time and probably may well not receive a fair amount either. This is where blockchain technology plays an important role. Using blockchain-based platforms provides a fully decentralized, peer-to-peer network with a distributed ledger to store the data related to the royalties and rights of the songs when streamed or downloaded. Based on the impact of online streaming of music in India and some companies that have already implemented blockchain for music industries from across the globe, we proposed a blockchain-based system for the Indian music industry in this chapter. A new cryptocurrency term, "Mol", was also proposed as a part of the system.

9.17 FUTURE SCOPE

In future, the platform developed can be used as a medium for upcoming artists to interact with their users; they can ask their fans to choose a certain genre in which the artist must create his or her next song or album. In other words, it would be like a social media platform for artists to interact with their fans (Chalmers et al., 2019). The features of the platform can be enhanced by making use of artificial intelligence techniques to identify the correct recipient of the royalty for a given song or music, making the platform more reliable and transparent (Owen and O'Dair, 2020). This would ensure more royalties for the artist, as the level of appreciation by the audience would be higher.

REFERENCES

An introduction to the music industry: Part 1. *Horus Music*, 31 Aug. 2020, www.horusmusic.global/an-introduction-to-the-music-industry-part-1

Arcos, L. C. (2018). The blockchain technology on the music industry. *Brazilian Journal of Operations & Production Management, 15*(3), 439–443.

Chalmers, D., Matthews, R., & Hyslop, A. (2019). Blockchain as an external enabler of new venture ideas: Digital entrepreneurs and the disintermediation of the global music industry. *Journal of Business Research.*

Daley, Sam. (n.d.a). 15 Examples of how blockchain is reviving healthcare. *Built In*, 25 Mar. 2020, www.builtin.com/blockchain/blockchain-healthcare-applications-companies

Daley, Sam. (n.d.b). 25 Blockchain applications & Real-world use cases disrupting the status quo. *Built In*, 25 Mar. 2020, www.builtin.com/blockchain/blockchain-applications

De León, I. L., & Gupta, R. (2017). The impact of digital innovation and blockchain on the music industry. Inter-American Development Bank. (Nov 2017). R&D Management Conference 2018 At: Milan, Italy. Available online: https://publications. iadb. org/en/impact-digital-innovation-and-blockchain-music- industry (accessed on 23 June 2020).

Drescher, D. (2017). *Blockchain basics* (vol. 276). Berkeley, CA: Apress.

How much data is on the internet? l The big data facts update 2020. *NodeGraph*, 28 Sept. 2020, www.nodegraph.se/how-much-data-is-on-the-internet/

Kim, A., & Kim, M. (2020, October). A study on blockchain-based music distribution framework: Focusing on copyright protection. In 2020 International Conference on Information and Communication Technology Convergence (ICTC) (pp. 1921–1925). IEEE, 21-23 Oct. 2020, Jeju, Korea. (South). DOI: https://doi.org/10.1109/ICTC49870.2020.9289184.

Mediachain: Documentation. *MediaChain*, Accessed 17 Dec. 2020. docs.mediachain.io

Moreira, M. A. B. R. (2020). *Innovating in the music industry: Blockchain, Streaming & Revenue Capture* (Doctoral dissertation).

Owen, R., & O'Dair, M. (2020). How blockchain technology can monetize new music ventures: an examination of new business models. *The Journal of Risk Finance, 21*(4), 333–353. https://doi.org/10.1108/JRF-03-2020-0053

Preethi, D., Khare, N., & Tripathy, B. (2020). Security and privacy issues in blockchain technology. *Blockchain Technology and the Internet of Things: Challenges and Applications in Bitcoin and Security, 245*, 236–255

Rosic, Ameer. What is blockchain technology? A step-by-step guide for beginners. *Blockgeeks*, 25 Nov. 2020, www.blockgeeks.com/guides/what-is-blockchain-technology

Shrestha, B., Halgamuge, M. N., & Treiblmaier, H. (2020). Using blockchain for online multimedia management: Characteristics of existing platforms. In *Blockchain and Distributed Ledger Technology Use Cases* (pp. 289–303). Cham: Springer.

Sitonio, C., & Nucciarelli, A. (2018) The Impact of Blockchain on the Music Industry. R&D Management Conference 2018, R&Designing Innovation: Transformational Challenges for Organizations and Society, pp. 1-13, June 30th -July 4th, 2018, Milan, Italy.https://doi.org/10.1108/JRF-03-2020-0053.

Statista. "Digital music: India l statista market forecast." *Statista*, www.statista.com/outlook/202/119/digital-music/india#market-users. Accessed 17 Dec. 2020.

Tabora, Vincent. A Decomposition of the Bitcoin Block Header. *Data Driven Investor*, 21 Nov. 2019, www.datadriveninvestor.com/2019/11/21/a-decomposition-of-the-bitcoin-block-header/#

Ucaya. Soundcharts l Market intelligence for the music industry. *Soundcharts*, www.soundcharts.com/blog/india-music-market-overview#2-recording-industry. Accessed 17 Dec. 2020.

Vujičić, D., Jagodić, D., & Ranđić, S. (2018, March). Blockchain technology, bitcoin, and Ethereum: A brief overview. In 2018 17th International Symposium Infoteh-Jahorina (Infoteh) (pp. 1–6). IEEE 2018, East Sarajevo, Bosnia and Herzegovina, 21-23 March 2018. DOI:10.1109/INFOTEH.2018.8345547

Wang, S., Ouyang, L., Yuan, Y., Ni, X., Han, X., & Wang, F. Y. (2019). Blockchain-enabled smart contracts: Architecture, applications, and future trends. *IEEE Transactions on Systems, Man, and Cybernetics: Systems, 49*(11), 2266–2277.

10 Blockchain Technology
Myths, Realities and Future

Ashish K Sharma, Durgesh M Sharma, Neha Purohit, Sangita A. Sharma and Atiya Khan

CONTENTS

DOI: 10.1201/9781003138082-10

10.1 INTRODUCTION

Blockchain is a distributed software network that works as both a digital logger and a tool that securely transfers assets without a third party. It refers to a chain of blocks that holds information. A Blockchain gathers information in clusters called blocks that contain sets of information. These blocks have storage capacities and when filled, are chained onto the blocks filled earlier, creating a sequence of data called a Blockchain. Blockchain timestamps digital documents, preventing them from being backdated or tampered with (Guru99, 2020). A Blockchain network allows almost anything from currencies to land deeds to votes to be tokenized, stored and replaced. Blockchain provides a quicker, more competent way for businesses to transfer, receive and track orders through secure data. This technology is poised to change nearly every facet of our digital lives. Many innovative uses of Blockchain technology have appeared recently. The most common involves the safe transfer of items such as cash, property, contracts, etc.

By evading censorship, Blockchain promises to make our systems more equitable. In traditional or conventional systems, the clearing control systems are centralized, and thus, there was a high probability of security risks. Various traditional technologies were introduced as security measures, such as cryptography, number generator, physical unclonable functions, etc. These security measures failed to counteract the effects of stealthy (secretive) attacks during the runtime or testing phase as well as the effects of uncertainties during real-time cases. However, in Blockchain, the blocks are distributed and decentralized in the whole network, which makes it more secure than the traditional systems. And if properly implemented, it could make our systems more reliable and secure. Blockchain spreads its processes across a network of computers, allowing Bitcoin and other cryptocurrencies to function without a central authority and thereby decreasing risk and removing the processing and transaction fees.

10.2 BLOCKCHAIN

Blockchain is a way to store, share and securely distribute data. Blockchain is a digital record file that preserves online transactions safely. In general terms, a Blockchain is best thought of as a shared digital record on which each transaction is logged. It can be described as an archive of fixed sequence data distributed and managed by a device or computer group. It can also be termed an irreversible timestamped sequence of data archives, which is decentralized and administered by a device or a computer group. The technique timestamps the digital documents, not allowing anyone to backdate them or tamper with them. Blockchain maintains constantly increasing data logs. It is in dispersed form, i.e., no main computer possesses the complete sequence. Instead, the contributing links have a duplicate of the sequence. The main difference between a Blockchain and a typical database is the method of structuring data. A database uses a table structure for data, while a Blockchain structures its data into blocks or chunks that are attached together. It gathers information in clusters called blocks that contain information sets. Blocks contain definite repository volumes and when filled, are attached to the previously filled block, making a data chain called the Blockchain. The essential idea when discussing Blockchain is that it is open source, meaning that anybody with a computer and adequate free storage space can download it. The objective is to resolve the issues of duplicate records without making use of a central server. It focuses on safe transfer of things such as money, assets, bonds, agreements, etc. without requiring an arbitrator such as a government or bank. Blockchains use cryptography to secure and protect transactions. Once data is embedded within Blockchain, it is very difficult to modify it. It is a software protocol (similar to SMTP email). But, distributed networks comprise a number of computers in different sites that share a data connection; there is no direct data connection among them (Guru99, 2020).

Blockchain is distributed, i.e. there is no centralized control that can authorize the transactions or rule set to approve or accept transactions. This signifies that an enormous amount of faith is needed from every member in the network system; they have to come to a consensus to approve dealings. Blockchains are distributed by nature; in short, there is no solitary main computer controlling the whole chain. Rather, any computer connected to the Internet can become an important node of the network, known as a "terminal". In order for transactions to be acknowledged, a certain number of members in the system network have to come to an agreement that a transaction has been carried out. The exact method by which this occurs differs between different Blockchains. The basic benefits of Blockchain technology are decentralization, immutability, security and transparency. With its distributed and unconfined nature, Blockchain technology can lead to new prospects and advantage businesses through better transparency, improved security and easier tracking. The great benefit of Blockchain is that it is public. Everybody contributing can see the blocks and the transactions kept in them. This does not signify that everybody can see the details of the transactions, though; those are secured by your private key. Most importantly, it's safe. The database can be extended, and the preceding logs are unchangeable (because it is very expensive if somebody needs to modify preceding documentation). Blockchain does not need the Internet. It is also known as a meta-technology, as it influences additional technologies, including databases, application software, other compatible computers, etc. (Brainbridge, 2019).

There are two features involved in Blockchain:

- Members of the system perform the transaction activities.
- Transactions are noted down by blocks, and it is then confirmed that they are in the right order and have not been altered or damaged.

On a certain network, all transactions that have ever been carried out are noted on the Blockchain, creating immutable evidence. This transaction evidence is kept as digital recordings known as blocks. Precisely what is required to authenticate transactions differs between Blockchains, but typically, this information contains primary transaction data like the cost, timestamp and operations. Every transaction block is cryptographically "secured or locked" to the preceding block because it holds a cryptographic hash. These aids keep a log of records of the sequence in which transactions happened, avoiding hacking and fraud. When anyone needs to add a transaction to the sequence of the chain, all the members in the block network approve it. This is done by introducing an algorithm or a procedure to validate its legitimacy. What precisely is meant by "authentic" is decided by the Blockchain system and generally varies among systems. After that, it depends on members' consensus to approve the transaction as correct and legitimate. Later on, a group of all accepted transactions is tied up in a set of blocks and then passed to all the nodes of the system network. These, in order, authenticate the fresh block. Every consecutive block holds a hash that represents the distinctive fingerprint of the preceding block. This confirms the need for the data to be altered, providing a slab of timestamping that eliminates several stages of human inspection and ensures smooth transactions. However, it isn't yet the panacea that some trust it to be (Banafa, 2018). In short, a Blockchain is a collection of data blocks associated by cryptographic tools so as to make it impractical to modify the data or content of a single block without intervening with all the others. In that digital record, information is kept in a network of distributed system nodes, and all logged transactions are crystal clear to each associate of the system network (Kosmarski, 2020).

10.2.1 Blockchain Architecture

A Blockchain architecture consists of various parameters that can provide security for the secure transfer of money, property, etc. without involvement of a third party. The architecture consists of elements like computers or nodes that are used for handling the transactions done on the records, information or systems. The Blockchain architecture involves the following components, which are shown in Figure 10.1:

- Transactions.
- Blocks.
- Peer-to-peer (P2P) network.
- Consensus algorithm.

10.2.1.1 Transactions

Transactions form the most minuscule construction blocks of a Blockchain system. They customarily consist of a receiver address, sender address and value. A standard credit card can be considered a good example of a transaction. The holder transfers the value by digitally signing the hash engendered by integrating the anterior transaction and the public key of the beneficiary. The transaction is then publicly promulgated to the network, all the nodes independently hold their own facsimile of the Blockchain, and the currently known "state" is calculated by processing each transaction as it appears in the Blockchain. They are distributed to each node in the form of a block. As the transactions are distributed throughout the network, each node is independently verifying the transactions that are in process.

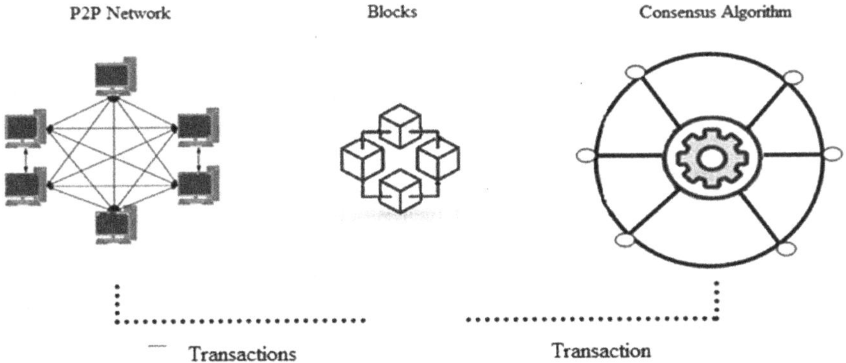

FIGURE 10.1 Blockchain architecture.

10.2.1.2 Block

The data that is stored inside a block of a Blockchain, which contains information in the form of a "chain of blocks", depends on the type of Blockchain. The first block in the chain is called the Genesis block. Each opposing block in the chain is connected to a new block. The information in the block is categorized as block header and transactions. A function of the block header is to verify the validity of a block. Each block has:

1. Data.
2. Hash.
3. Hash of the previous block.

The blocks in the Blockchain contains the hashes of the antecedent block, which provides a feature of security in Blockchain. The blocks in the Blockchain are created by miners. The process in mining is to create a block that is valid, such that it is accepted by the rest of the network. The role of nodes is to handle the pending transactions in the network, verify that they are cryptographically accurate and wrap them into the blocks to be stored on the Blockchain.

10.2.1.3 P2P Network

A peer-to-peer (P2P) network indicates that there is no centralized node and provides more security because there is no single point of attack or failure as there would be in a centralized network. Blockchain architecture uses a P2P network, which is based on the IP protocol. Each and every single node in a network keeps a local copy of a Blockchain, and the decentralization of the Blockchain architecture is built on the P2P network. To provide more security among the blocks and nodes, a Blockchain uses a distributed P2P network, which everyone is allowed to join easily and will have access to a full copy of the Blockchain. When any user creates a new block, it is sent to all users (nodes) in the network, each node verifies the block, and after full verification, each node in the network adds this block to their Blockchain. In this process, all nodes in the network are compatible and agree whether or not the block is valid. When any user creates a new block, it is sent to all the users (nodes) on the network, each node verifies the block, and after complete verification, each node in a network adds this block to their Blockchain. In this process, all the nodes in a network create a consensus, i.e. they agree which blocks are valid and which are not. Blocks that have been tampered with are rejected in the network.

10.2.1.4 Consensus Algorithm

The way the copies are synchronized in a Blockchain is because of a consensus algorithm. The algorithm makes sure that every individual party that has a local copy of a Blockchain should be consistent with the others, and the copy should be the most recent one. The compatibility algorithm is at the heart of Blockchain architecture. Various algorithms for compliance are Proof-of-Work (POW), Proof-of-Stake (POS) and Simplified Byzantine Fault Tolerance (SBFT), which is used for implementation.

10.2.2 Types of Blockchain

An understanding of the types of Blockchain networks is vital for achieving background knowledge of Blockchain and how to implement it for cryptocurrency (Sharma, 2020).

10.2.2.1 Public

This is a Blockchain that has an open network where the data is accessible to all, and thus, anybody can contribute to the agreement procedure. It is permissionless, and thus, anyone can interpret, read and transcribe data on the Blockchain. It is distributed and trustless and doesn't have a sole organization that regulates the network. The data here is safe, as it becomes impossible to change or modify data once the data is authorized on the Blockchain. Bitcoin and Ethereum are popular models of this type of Blockchain.

10.2.2.2 Federated

This Blockchain doesn't permit everybody to contribute to the consensus process. Only a partial number of nodes are authorized to do so. Access to the Blockchain is either public or limited to members. This is suitable for smaller teams. Hyperledger and Corda are models of a consortium Blockchain.

10.2.2.3 Private

A private Blockchain is an invitation-only Blockchain. This is administered by a single entity. These are typically used in a business. Only particular associates are permitted to access it and perform transactions. This is an authorized Blockchain. It is based on access controls that specify the public that may contribute to the system network. Some entities regulate the network, and this gives rise to trust on third parties to perform transactions. Here, only the actors contributing to a transaction will possess information about it, while others will be unable to access it. It is more vulnerable to threats, hacks, and data breaks. Linux Foundation Hyperledger Fabric is the best illustration of private Blockchain. Public Blockchain appears to be the prominent choice; it is implemented in most use cases, since it is not restricted by limited access. Amalgamating public and private Blockchains signifies feasible solutions for companies.

10.2.3 Benefits of Blockchain

10.2.3.1 Better Transparency

Blockchain is based on a distributed ledger, with all network participants sharing the same text, which can only be updated accordingly. To modify a single transaction record would require a change in all succeeding records and the collusion of the entire network; this gives rise to better transparency.

10.2.3.2 Improved Security

Blockchain offers improved security, as transactions need to be agreed upon prior to being recorded. Once a transaction is approved, it is encrypted and linked to the previous transaction. Also, information is stored across a network of computers rather than on a single server, making it complex for hackers to compromise the transaction data.

10.2.3.3 Better Efficiency and Speed

Blockchain greatly assists in streamlining and automating the trading processes, and thus, the transactions can be completed quickly and more proficiently.

10.2.3.4 Cost Effective

Reducing cost is a priority for businesses, and since Blockchain removes the need for any third parties or middlemen to make guarantees in a trade, it proves more cost effective.

10.3 LITERATURE SURVEY

Decision making forms an important basis for businesses. The significance of decision making and its use have been demonstrated by (Sharma et al., 2009, 2012; Purohit & Sharma, 2015; Sharma & Khandait, 2016; Sharma & Khandait, 2017). Distributed ledger technology has made strategic decision making more challenging than ever. Blockchain has also made major contributions to decision making. In his article of 21 August 2018, Braiman (2018) has discussed in detail how Blockchain could transform decision making. Tapus and Manolache (2019) have integrated decision making using the Blockchain. Blockchain technology has been applied and implemented successfully by different authors in varied areas such as the Internet of Things (IoT), health, business and food chain supply, data management, integrity verification, etc. Reyna et al. (2018) investigated challenges in Blockchain IoT implementation and surveyed the most applicable work so as to examine how Blockchain can possibly make a better IoT. Huh et al. (2017) built a Blockchain-based IoT system that can regulate and organize IoT units on the Ethereum platform. Agbo et al. (2019) presented a systematic review to illustrate how Blockchain technology can be successfully utilized in healthcare. The review shows that a number of studies have proposed different use cases for the application of blockchain in healthcare; however, there is a lack of adequate prototype implementations and studies to characterize the effectiveness of these proposed use cases. . McGhin et al. (2019) discussed the research challenges and prospects involved in Blockchain in healthcare applications and also explored various research agendas, such as in the domain of patient identity validation in an emergency situation. Research remains to be done on a real-world patient dataset so as to verify its outcomes and act accordingly. Keogh et al. (2020) presented a systematic review on the possibilities and challenges of Blockchain and GS1 standards in the food supply chain and observed that the amalgamation of GS1 and Blockchain technology would provide a rational structure for a business-based method for transforming the food supply chain and would also improve the food traceability system. Rogerson and Parry (2020) investigated using case studies and found that Blockchain has moved on from cryptocurrencies and is being implemented to enhance transparency and faith in supply chains, their restrictions and potential influence. Chen et al. (2019) designed an employee big data management system based on Blockchain that solves the Blockchain information repetition issue, analyzed the shortcomings of deploying the proposed system with Blockchain, such as the viability of combining, and validated their ideas. Tian et al. (2019) proposed a novel method of establishing a distributed key to secure the confidentiality of healthcare information, and the simulation showed that the proposed module performed effectively. Liu et al. (2017) proposed a Blockchain-based protocols framework for IoT data, (Data Owner Apps) DOA to CSS-N (Cloud Storage Service), (Data Consumer Apps) DCAs to CSS-Y, and DOA to CSS-Y, and DCAs to CSS-N for maintaining data integrity, and found that the proposed structure could aid the integrity validation of data blocks. Wei et al. (2020) proposed a Blockchain data-based cloud data integrity protection mechanism by deploying a distributed virtual machine agent framework in the cloud with the help of mobile agent technology. Forecasting is one important area that has been extensively used by researchers in various fields, one being forecasting of customer requirements in quality function, as illustrated by Purohit and Sharma (2017).

10.4 MYTHS AND REALITIES OR FACTS

10.4.1 Myth 1: Blockchain Is a Magical Database in the Cloud or Blockchain Does Not Store Any Type of Document

One popular myth is that Blockchain is some magical database in the cloud. It is a sequential list of transaction records, which can be termed a flat file, i.e., a list that appends entries that are never deleted, and which is duplicated in each terminal in a P2P network. It provides the distributed ledger, i.e. the "proof-of-existence" that certifies or indicates the presence of a definite record but not the record itself. Most people believe that it is possible to store documents on a Blockchain, but the reality is totally different. In reality, it only contains the information that a specific document, such as a spreadsheet, a pdf, etc., exists. These types of documents are stored in data lakes and can be accessed only by the owner of the document. This myth is identical to the fact of cloud storage, which is similarly something that exists but cannot be described. The cloud does not have any physical drive for storing information, and the same goes for Blockchain; there is no physical storage device for Blockchain transactions (thenextweb, 2018).

10.4.2 Reality

Blockchain does not store any document. It stores the records of transactions done; whether failed or successful transactions, they are never deleted. In other words, it stores the work done by the owner from start to end. In reality, as one of the myths implies that it is cloud based, a Blockchain can be downloaded and run on Internet-enabled computers for operational functions.

10.4.3 Myth 2: Blockchain Will Transform the World

It is usually said of Blockchain that it is such a powerful and revolutionary technology that it is going to be widely accepted by companies worldwide to reap rich benefits. However, this is only partially true.

10.4.4 Reality

In dealing with complex and technical transactions, Blockchain can play a crucial role. It can deal with transactions like identity of a person or in trade finance. The application of Blockchain in trade finance can help with cost reduction and transaction speed. In addition, dealing with and maintaining a traditional ledger can be inefficient in many cases. While Blockchain mitigate the risk of a fraudster tampering with the ledger, it does not eradicate the threat of fraud online and it still raises questions over confidentiality. So, it should be noted that Blockchain has the potential to change the world but has still not achieved its objective.

10.4.5 Myth 3: Blockchain Is Free

One of the most usual myths is that Blockchain is either free or very cost effective. But the fact is that this is not true.

10.4.6 Reality

Blockchain is not cost efficient, nor is it cost effective to run the system. For processing, it needs multiple computers for solving complex problems to havc the same opinion upon the final rigid results which is called a Single version of Truth (SVT). Each block in a Blockchain employs massive computing ability to resolve a difficult as well as complex problem, and all the services need to be paid for. Another fact of Blockchain is Bitcoin mining, which requires very powerful hardware for processing and consumes in the order of terawatt-hours of electricity, which indicates high cost. Hence, it cannot be said that Blockchain is free.

10.4.7 Myth 4: There Is Only One Blockchain

It is a common belief that there is only one Blockchain.

10.4.8 Reality

The fact is that Blockchain has many different technologies, including public and private versions, open and closed source, and many more. Considering a Bitcoin as a public global Blockchain on the basis of its introduction can confuse any beginner. So, on the basis of facts, private and hybrid Blockchain can also be considered for different complex problems. Many other distributed ledger technologies, like

Ethereum, Hyperledger, Corda, and IBM and Microsoft's Blockchain-as-a-service, are examples of different types of Blockchain in the present world.

10.4.9 Myth 5: Blockchain Can Be Used for Anything and Everything, or Blockchains Can Be Used Everywhere

This highlights the myth that Blockchain is ubiquitous and can be used for anything and everything; Blockchains can be used everywhere. However, in its present form, it has certain restrictions, which defy the above point.

10.4.10 Reality

This is connected to the assumption that in future, Blockchain and smart contracts will substitute for many assets, like money, lawyers and other arbitration bodies. Blockchain represents resource management and contracts (depending on the algorithm) that cannot be changed. In other words, whenever resources are involved, it is hard to resolve disputes with the software.

10.4.11 Myth 6: Blockchain Can Be the Backbone of a Global Economy

The myth about Blockchain is that it can be an integral part of the global economy.

10.4.12 Reality

Blockchain is independent; no agencies, national or corporate, are involved in controlling it. Hence, it is assumed to be private. Blockchain acts as a backbone for the various encrypted, trusted cryptocurrencies: Bitcoin Blockchain, etc. In a survey, the Gartner report asserted that Blockchain is similar in scale to the National Association of Securities Dealers Automated Quotations exchange Network (NASDAQ). A Blockchain network can be converted into a financial network for a global economy.

10.4.13 Myth 7: The Blockchain Ledger Is Locked and Unchangeable/Unable to Modify the Data Block Once Created, or Blockchain Data Cannot Be Changed Once Updated

Another myth about Blockchain is that the Blockchain ledger is locked and unchangeable, i.e. Blockchain data cannot be changed once updated because of inability to modify the data block once it is created. Big transaction databases such as bank records are usually private and linked to particular financial organizations. In Blockchain, the code happens to be public, and transactions are certifiable. Also, the network is cryptographically protected. This ensures that fraudulent transactions like double spends are disallowed by the network, preventing fraud. Moreover, redrafting important transactions is in no way in the financial interest of members, as mining the chain offers financial motivation in the form of Bitcoin.

10.4.14 Reality

The illusion of immutability stems from its append-only data structure. However, the fact is that the data block that has been created can be modified and reversed in special circumstances by the participants. Participants can be state actors, individuals or a company. Even changes in the distributed ledger can be possible if the network's participants decide this themselves. The vital participants are miners, who manage the entire work. So, if a majority of miners decide to do so, they can manipulate or modify the distributed ledger network.

10.4.15 Myth 8: Blockchain Records Can Never Be Hacked or Changed/Blockchain Technology is 100% Trustworthy, or Blockchains Are Truth Machines

Some myths about Blockchain focus on its security aspects, such as: Blockchain records can never be hacked or changed, Blockchain technology is truly trustworthy, or Blockchains are Truth Machines. The common myth is that Blockchains are invulnerable to outside attacks.

10.4.16 Reality

The fact is that Blockchains can only provide a way of catching unauthorized modifications for applications that are designed on top of them. Moreover, cryptography technology – the technology behind Blockchain – needs to be trusted rather than trusting the Blockchain. The cryptography method is embedded into Blockchain, which enables it to provide the security level. However, Blockchain fails to check out a trust level that arrives from external sources. For example, the Bitcoin-encrypted network can be decrypted and hacked, which may lead to destroying its entire Blockchain system. The fact is that Blockchain cannot verify the health of input data and is best suited if the asset being transferred is part of a network, e.g. Bitcoins.

10.4.17 Myth 9: Blockchain Can Only Be Employed in the Financial Area

When it comes to Blockchain application areas, the financial sector inherently grabs the first place, the reason being Blockchain's first application, the Bitcoin cryptocurrency, which influenced this area. Major challenges posed by Blockchain in the financial world have led foreign banks such as Goldman Sachs and Barclays to invest heavily in it. This has created the myth that it can only be used in the financial sector.

10.4.18 Reality

In reality, Blockchain can be successfully utilized in varied areas. Application areas include real estate, healthcare, insurance, supply chain or even at an individual scale to build a digital identity. People can keep evidence of the availability of medical information in Blockchain and give access to pharmaceutical companies to make money.

10.4.19 Myth 10: Blockchain Is Bitcoin and Bitcoin Is Blockchain

There is a common misconception that Blockchain and Bitcoin are one and the same thing. Bitcoin was the first application of Blockchain. Moreover, Bitcoin is more famous than Blockchain, and thus, individuals become confused between the two and mistakenly use Bitcoin to mean Blockchain. However, this is just a myth. This myth is extensive, since many individuals assume that Bitcoin Blockchain is the only Blockchain and that the two can be substituted for each other.

10.4.20 Reality

Bitcoin and Blockchain are not the same, but they are very closely related. Blockchain is a technology that allows P2P transactions to be written in a spreadsheet distributed across the network. In Blockchain, everything is transparent and permanent. Changing or removing a transaction from the ledger is not permitted to anyone. On the other hand, Bitcoin is a cryptocurrency that allows direct e-payment between two parties without any intermediary or third party such as a financial institution. Bitcoins are generated and loaded in a virtual wallet. Blockchain has several additional potential usages in all business areas, comprising supply chain, insurance, identity verification and medical (Hall, 2019).

10.4.21 Myth 11: Blockchain Is Developed for Big Companies/Commercial Communications Only

Considering the strength of Blockchain and going by the experts in Blockchain, it is believed that Blockchain will transform the world as well as the worldwide economy, just as dot-coms did before the 1990s. Therefore, it is open to large companies and restricted to business interactions. The myth that Blockchains are only for big businesses may stem from the fact that there are indeed many large companies working on Blockchain projects.

10.4.22 Reality

Blockchain clearly defies the myth that it is designed for business interactions only. It is accessible to everybody ubiquitously. All anyone needs is an Internet to employ the Blockchain. Thus, it can be clearly seen that many individuals globally will be able to interrelate together. Blockchain is surely not meant for big businesses only. Non-corporate users or small companies can also employ Blockchain. It is designed in such a way that it can be scaled to fit the needs of the user, thus paving the way for a wide variety of individuals, groups, businesses and non-business users to benefit from it (thenextweb, 2018).

10.4.23 Myth 12: Smart Contracts Have a Similar Legal Value to Regular Contracts

It has become a common myth that smart contracts hold the same legal value as standard contracts.

10.4.24 Reality

Smart contracts represent the lines of code that perform activities automatically when specific circumstances are encountered. Thus, they do not count as fixed contracts from a legal viewpoint. Nevertheless, they are utilized as evidence to identify whether a particular job has been done. Albeit smart contracts have uncertain legal value, they act as strong tools particularly when merged with the IoT.

10.4.25 Myth 13: Blockchain Can Store Documents

Blockchains are not about data storage, and they do not stock any physical documents or data such as text or pdfs.

10.4.26 Reality

In reality, a Blockchain doesn't store documents; rather, it only contains the information that a specific document exists. Instead of storing a document, it stores POW only in the distributed ledger, which records all transactions, either failed or successful, and these are never deleted. Data like documents or spreadsheets are kept in data lakes, and the right of these resides only with the holder. Blockchain offers proof of existence; the distributed record comprises a code that confirms the presence of a definite file and document.

10.4.27 Myth 14: Blockchains Are Intended Only to Aid Money Management

This myth is driven by the fact that Blockchains are suitable for financial transactions. However, Blockchain is not restricted to managing only financial transactions.

10.4.28 Reality

The use of Blockchains can be witnessed in managing other transactions, such as accelerating transactions and transaction agreements, improving transactions and cross-border transfers, and building perceived asset-to-end purchases to quicken the dealing and agreement of transactions, to improve cross-border trade-payments and transfers, and to build a supply chain that is observable from end to end. A Blockchain can increase clarity on every side of a transaction – from imports roaming over a supply chain to validation of qualification as well as grades. In addition, it can validate the testimonials of a career aspirant (Hall, 2019).

10.4.29 Myth 15: Blockchains Consume High Volumes of Energy, Increasing Costs

That Blockchains consume high volumes of energy, driving up costs, can be considered to be true to some extent. However, it doesn't apply to all Blockchain transactions and depends on the mining being used.

10.4.30 Reality

Depending upon the way the Blockchains are used, such as permissioned or permissionless, this myth can be suitably evaluated. Mining done through permissionless Blockchains inflates costs. However, this is false for permissioned Blockchains, as they do not normally include cryptocurrency mining. The administrator of reliable members designs the rubrics to authenticate information in the system on the chain. These Blockchains are more cost effective (Hall, 2019).

10.5 CONCLUSION

Blockchain has emerged as an exciting new technology that is being widely used nowadays because it provides better transparency, better traceability, higher security, better efficiency and speed, and is cost effective. From supply chain to manpower resources, Blockchain influences all the sectors of industries it touches. Blockchain implements applications in a decentralized and secure way, ensuring certainty. This has led to the widespread adoption of Blockchain in a trustless society. However, exciting new technologies often create disruption in the marketplace; disruption breeds misunderstanding, and Blockchain technology is no exception. Among several technological developments in the current era, Blockchain has been an eminent challenger. So, the lack of transparency for understanding Blockchain has resulted in the generation of several Blockchain myths over time, which may hinder people from using or adopting it. Myths about distributed ledger technology and its working mechanism – Blockchain is the only Bitcoin, is secure, cannot be manipulated, etc. – pose the threat that industries will avoid exploiting its extensive potential to influence constructive development. This generates the pressing need to explore the various myths and realities related to Blockchain and separate the myths from realities. To this end, this chapter has discussed Blockchain technology and its concepts and has presented the various myths and reality associated with Blockchain. The chapter has collected and reviewed the contents of various research papers and online sources to serve this purpose. It has attempted to sort or separate the myths from the reality, highlighting the fact that using Blockchain is not restricted to only some domains. The chapter has clearly unravelled some of the commonly assumed misconceptions and removed misperceptions, thereby providing users with a better understanding of Blockchain. It is hoped that the chapter will greatly assist users in understanding the concepts, knowing the myths and reality associated with Blockchain, and gaining insight into a few of the common misperceptions. It is also hoped that the chapter will provide a strong foundation to researchers who wish to pursue research in this domain.

10.6 FUTURE SCOPE/DIRECTIONS

The chapter in its present state has clearly demonstrated Blockchain technology and various myths and realities associated with it so as to provide users with a better understanding of Blockchain. Blockchain, being such a powerful technology, has a broader future scope for implementation in various areas that have remained untouched, like digital advertising, managing world trade, forecasting, the IoT, and networking and cloud storage. Blockchain technology could be used in many sectors in the future as well as in government programs, as these systems are slow, congested and potentially corrupt. Implementing Blockchain technology in a government system can make its operations much safer and more efficient. Supply chain management is also one vital area where Blockchain can be used to make transactions more effective and transparent and reduce human error and time delays. Future implementation will involve exploring its potential in these areas and also exploring any myths that may not have been included in this study.

REFERENCES

13 Blockchain Myths Everyone Believes. Brainbridge. https://www.brainbridge.be/news/13- Blockchain-myths-everyone-believes

Agbo, C. C., Mahmoud, Q. H., & Eklund, J. M. (2019, June). Blockchain technology in healthcare: A systematic review. *Healthcare*, *7*(2), 56. Multidisciplinary Digital Publishing Institute.

Banafa, A. (2018). *12 Myths about blockchain technology.* Industrial Automation India. https://www.industrialautomationindia.in/articleitm/2670/12-Myths-about-Blockchain Technology/articles

Binary District Journal (2018). *The 13 most common blockchain myths explained.* The Next Web. https://thenextweb.com/syndication/2018/03/11/13-common-blockchain-myths-explained/

Blockchain tutorial: Learn blockchain technology (examples). Guru99. https://www.guru99.com/blockchain-tutorial.html

Chen, J., Lv, Z., & Song, H. (2019). Design of personnel big data management system based on blockchain. *Future Generation Computer Systems*, *101*, 1122–1129.

Hall, M. (2019, February 25). *Why Businesses Need Blockchain: Myth vs. Reality.* Oracle Blockchain Blog. https://blogs.oracle.com/blockchain/why-businesses-need-blockchain:-myth-vs-reality

Huh, S., Cho, S., & Kim, S. (2017, February). Managing IoT devices using blockchain platform. In 2017 19th International Conference on Advanced Communication Technology (ICACT) (pp. 464–467), IEEE, Phoenix Park, PyeongChang, Korea.

Keogh, J. G., Rejeb, A., Khan, N., Dean, K., & Hand, K. J. (2020). Blockchain and GS1 standards in the food chain: A review of the possibilities and challenges. In Detwiler, D., (Ed.), *Building the Future of Food Safety Technology*, Elsevier.

Kosmarski, A. (2020). Blockchain adoption in academia: Promises and challenges. *Journal of Open Innovation: Technology, Market, and Complexity*, *6*(4), 117.

Liu, B., Yu, X. L., Chen, S., Xu, X., & Zhu, L. (2017, June). Blockchain based data integrity service framework for IoT data. In 2017 IEEE International Conference on Web Services (ICWS) (pp. 468–475). IEEE.

McGhin, T., Choo, K. K. R., Liu, C. Z., & He, D. (2019). Blockchain in healthcare applications: Research challenges and opportunities. *Journal of Network and Computer Applications*, *135*, 62–75.

Purohit, S. K., & Sharma, A. K. (2015). Database design for data mining driven forecasting software tool for quality function deployment. *International Journal of Information Engineering & Electronic Business*, *7*(4), 39–50.

Purohit, S. K., & Sharma, A. K. (2017). Development of data mining driven software tool to forecast the customer requirement for quality function deployment. *International Journal of Business Analytics (IJBAN)*, *4*(1), 56–86.

Reyna, A., Martín, C., Chen, J., Soler, E., & Díaz, M. (2018). On blockchain and its integration with IoT. Challenges and opportunities. *Future Generation Computer Systems*, *88*, 173–190.

Rogerson, M., & Parry, G. C. (2020). Blockchain: Case studies in food supply chain visibility. *Supply Chain Management: An International Journal*, *25*(5), 601–614.

Sharma, A.K., & Khandait, S. (2016). A novel software tool to generate customer needs for effective design of online shopping websites. *International Journal of Information Technology and Computer Science*, *83*, 85–92.

Sharma, A.K., & Khandait, S. (2017). Development of a fuzzy integrated software tool to prioritise technical requirements for effective design of online shopping websites. *International Journal of Computational Systems Engineering*, *3*(1-2), 91–110.

Sharma, A. K., Mehta, I. C., & Sharma, J. R. (2009). Development of fuzzy integrated quality function deployment software: A conceptual analysis. *I-Manager*'s *Journal on Software Engineering*, *3*(3), 16.

Sharma, A. K., Sharma, J., & Mehta, I. C. (2012). A novel fuzzy integrated technical requirements prioritization software system for quality function deployment. *International Journal of Computers and Applications*, *34*(4), 241–248.

Sharma, T. (n.d.). *Public Vs. Private Blockchain: A Comprehensive Comparison*. Blockchain Council. https://www.blockchain-council.org/blockchain/public-vs-private-blockchain-a-comprehensive-comparison/

Tal, Braiman, (2018, August 22). *how Blockchain could transform decision making. Medium.* https://medium.com/@talbraiman/how-blockchain-could-transform-decision-making-6f5f6c0b9909

Tapus, N., & Manolache, M. A. (2019). Integrated decision making using the blockchain. *Procedia Computer Science*, *162*, 587–595.

Tian, H., He, J., & Ding, Y. (2019). Medical data management on blockchain with privacy. *Journal of Medical Systems*, *43*(2), 26.

Wei, P., Wang, D., Zhao, Y., Tyagi, S. K. S., & Kumar, N. (2020). Blockchain data-based cloud data integrity protection mechanism. *Future Generation Computer Systems*, *102*, 902–911.

11 Application and Challenges of "Blockchain Technology" in the Oil and Gas Industry

Adarsh Kumar Arya

CONTENTS

DOI: 10.1201/9781003138082-11

11.1 INTRODUCTION

Petroleum and natural gas are the primary energy sources that play a significant role in regulating a country's economic and social growth (Arya and Honwad, 2018b). Globally, 57% of all energy consumption comes from petroleum and gas exploitation. Over the next few decades, the petroleum and natural gas industry will thrive in the global energy sector. British Petroleum (BP) Energy Outlook 2019 states that the world is actively promoting new energy sources like wind, solar, biomass and hydrogen, but by 2040, half of the world's energy requirement will continue to be satisfied through petroleum and gas consumption only (BP Energy Outlook, 2019). Petroleum and gas reserves have played an essential part in the energy business. Typical exploration equipment, refineries and pipeline facilities have sophisticated facilities that must be carefully planned (World Economic Forum and Accenture, 2017; Fraser et al., 2018). However, the industry has a relatively old control system with poor performance, high cost and high-risk characteristics. Petroleum and gas companies have been prone to cyber threats. In the upstream market, machine monitoring has become complex. There were also data loss concerns. Inadequate evidence leads to poor decisions. There have been issues in handling duplicate data in the midstream sector, which has led to delayed transactions. In the downstream industry, integrity and security are significant challenges. Figure 11.1 shows some of the consequences of the lack of blockchain in the petroleum and gas sector.

These issues have triggered the petroleum and gas industry to search for a comparatively new technology that may improve management decisions. In these circumstances, blockchain technology has incredible benefits in the petroleum and gas sector. Figure 11.2 demonstrates that we can strengthen the existing processes inside the energy field by integrating blockchain technologies. In 2008, the invention of Bitcoin contributed to a surge in blockchain technology development (Midstream companies and oil and gas downturn Deloitte insights, 2019). As the petroleum and gas industry has faced massive challenges in protecting against the ever-looming cyber-attacks and cyber hacking risks, BP started looking into blockchain technology implementation. The move was an initial step towards adopting blockchain technologies in the petroleum sector. In 2018, a global blockchain survey conducted by Deloitte revealed that 61% of the petroleum and gas respondents understand that "blockchain is just a ledger and a program for financial assistance" (Deloitte global blockchain survey, 2018).

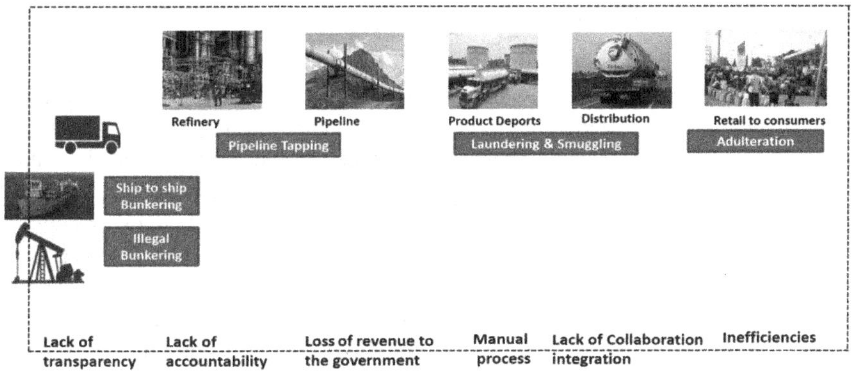

FIGURE 11.1 Consequences of lack of blockchain implementation to the petroleum and gas sector. (From Uba, S., *Blockchain adoption in the oil and gas supply chain*, 2021)

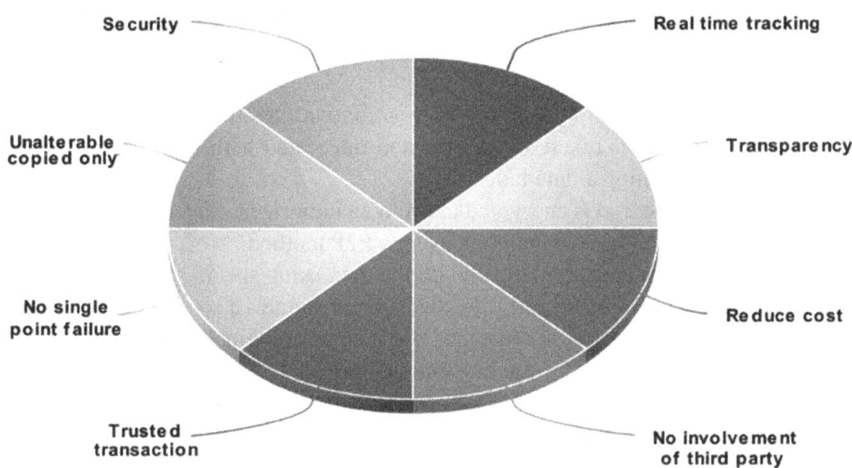

FIGURE 11.2 Benefits of integrating blockchain technology into the petroleum and gas industry.

Blockchain technology has been adopted by several industries for over a decade. However, the petroleum and gas sector has only just begun to adopt and research blockchain technology, revealing greater reservations about its application. This chapter aims to support the advancement of petroleum and gas blockchain technologies. Five key areas, blockchain fundamentals, applications, status, limitations and future directions, are discussed in detail in the following sections.

11.2 BLOCKCHAIN TECHNOLOGY FUNDAMENTALS

This section presents a detailed discussion of blockchain philosophy, attributes, consensus algorithms, encryption and model data algorithms.

11.2.1 Blockchain Philosophy

Explanations of blockchain in the literature are not entirely uniform. Practically, blockchain is like contract enforcement, which regulates transaction processing using transparency and protected rules to create blockchain peer-to-peer (P2P) network knowledge systems that cannot be forged or amended and are traceable. The main breakthrough in blockchain technologies is to disperse transfers to the users, not the central pool. P2P implies that computers are identical throughout every network node, and each node has the same network capacity but no centralized server. Both nodes exchange some services or knowledge using particular protocols. Transactions shall conform to major financial organizations' rules, and the transaction documents are primarily stored and managed by central authorities. With no other third-party interference, all transactions are processed and held privately for every person. Thus, the blockchain removes third-party control. However, with the absence of the central authority, it is impossible to validate the contract and maintain the ledger's credibility. This needs a proper verification method called a consensus algorithm (Zheng et al., 2017; Sikorski et al., 2017; Hamida et al., 2017; Sarmah, 2018). Figure 11.3 illustrates the algorithm and the methodology adopted in implementing blockchain technology.

Step 1: If an individual agrees to a transaction with another individual, then transaction data is taken as variable and integrated with trade simultaneously, thus creating a data block.

Step 2: The data stored is encrypted to avoid unauthorized usage and is further distributed to multiple computers using a P2P method.

Step 3: Network members now validate the data using specific algorithms. If the data is delivered correctly, the algorithm generates a correct hash value.

Step 4: After successful verification, the data is combined with the initial block, confirming the transaction's completion.

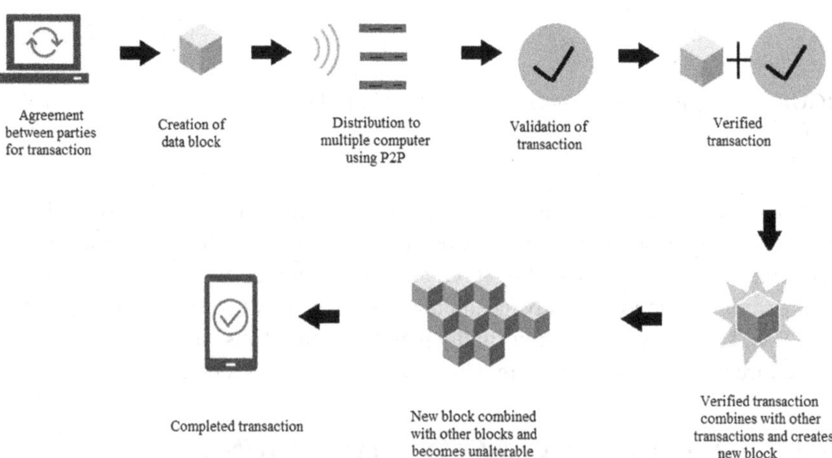

FIGURE 11.3 Methodology of Implementing blockchain technology.

11.2.2 Blockchain Key Attributes

Accountability, decentralization, immutability, anonymity and confidentiality are critical features of blockchain technology. Figure 11.4 reflects these key attributes. Decentralization ensures that the data is not centrally collected, processed and updated. Openness is the cornerstone of the trusted blockchain, as data recording and upgrading are open to nodes across the network. Providing a database archive for mobile devices makes the system simpler and more profitable. The performance and efficiency of the system are greatly improved using blockchain technology. Blockchain software rules implicitly determine whether node-to-node communication transactions are legitimate. The two parties make the transaction anonymous for blockchain structures, as the software rules determine whether the transactions are legitimate (Atlam and Wills, 2019; Sultan et al., 2018; Zambrano, 2018; Jansen, 2018; Abe et al., 2018; Hughes et al., 2018; Christidis and Devetsikiotis, 2016; Dorri et al., 2019).

11.2.3 Blockchain Infrastructure

Trusted and the public are the two groups of the blockchain infrastructure. Private and consortium are the two types of trusted blockchain. Currently, only commercial enterprises, government entities, or individuals are utilising private blockchain technology. The consortium rules recommend the amount of control a consortium participant should have over the blockchain and accounting records. It sets up a structured framework for data protection and safety concerns. It integrates private and state blockchain technologies. Nonetheless, only the blockchain would be vulnerable in a centralized network, and not the whole device's security would be

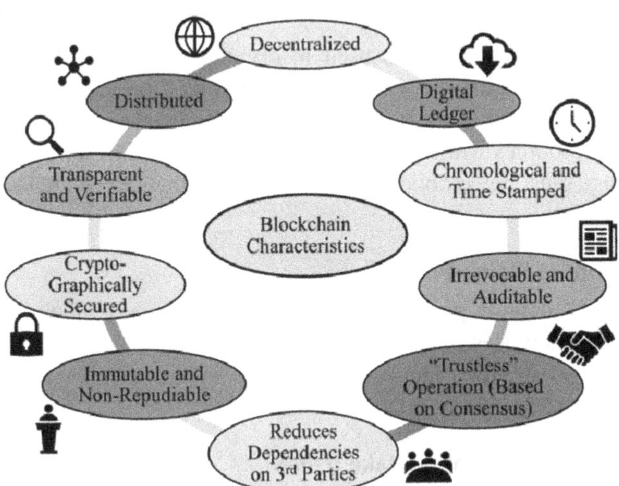

FIGURE 11.4 Blockchain key attributes. (From Puthal, D., et al., *IEEE Consumer Electronics Magazine*, 7, 18, 2018).

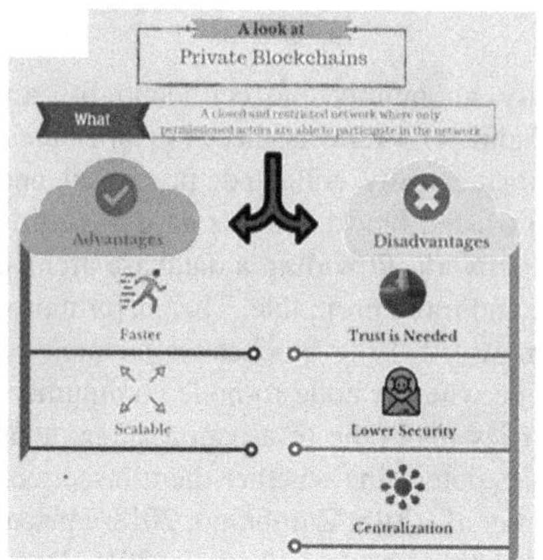

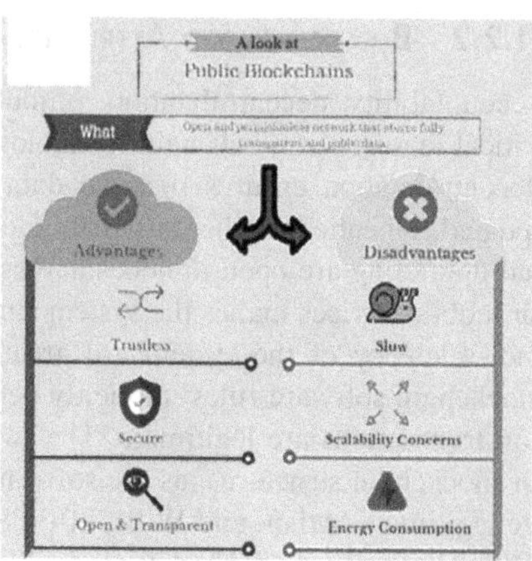

FIGURE 11.5 Private and public blockchains.

jeopardised.. Everybody in the world can freely use open blockchains. True decentralization is the mutual network ledger (Casino et al., 2019; Samaniego et al., 2016; Pilkington, 2016). Figure 11.5 illustrates the difference between public and private blockchains.

11.2.4 Consensus Algorithm

Consensus algorithms are essential innovations of blockchain. Blockchain consensus can be fault-tolerant and accurate. The consensus algorithm is explained in six consensus algorithms (Yuan and Wang, 2016).

11.2.4.1 Proof of Work (PoW)

The PoW consensus algorithm is used in cryptocurrencies such as Bitcoin and Litecoin. The original purpose of the organization was to remove spam in the network. The PoW algorithm's idea is to allow the participant to use computing power and economic resources while performing the PoW algorithm. The packets that the nodes exchange must be complicated and entail a mathematical challenge to determine their contents.

11.2.4.2 Proof of Stake (PoS)

A community of auditors performs an audit, documents any block of data on the blockchain, and then uses the web of confidence to relay votes about the block to the larger group of auditors. The PoS algorithm's key advantages include improved security and increased energy performance.

11.2.4.3 Delegated Proof of Stake (DPoS)

This algorithm is closer to PoW and PoS algorithms. Each elector can create several nodes that shape the democratic consensus. It efficiently preserves the entire device, its elements and its interdependencies.

11.2.4.4 Practical Byzantine Fault Tolerance (PBFT)

The classical Byzantine fault-tolerant method is no longer inefficient and the difficulty of implementing the algorithm has been minimized using PFBT algorithm. PBFT's security is ensured by all of the system's nodes. Three votes between nodes are the most important factor in determining the consensus. In a majority rule, each node has one vote, and each node's vote is represented by one node. As long as there are at least two-thirds of the (3a + 1) nodes in operation, the PBFT algorithm will continue to function correctly (a: Number of nodes).

11.2.4.5 Proof of Elapsed Time (PoET)

PoET is a lottery protocol that was built by Intel for trusted execution. It uses the Intel SGX platform. In PoET, the CPU reliability is measured by the time it takes for the hardware to respond to the environment. Generally, if the lowest latency is chosen, then the more CPUs are added to increase the system's resources.

11.2.4.6 Tendermint

A new algorithm, Tendermint, is being developed to compete with the PBFT algorithm. All it requires is a two-round vote. Just over two-thirds of verifiers propagate the same block transmit to the string during the same validation period. Often, the validator doesn't submit a block because the current provider isn't accessible or the network is sluggish. In Table 11.1, we present a comparison of various consensus algorithms.

11.2.4.7 Cryptography

The hashing algorithm is the most widely used crypto block string algorithm and is an essential participant in blockchain technology. It compresses arbitrarily long messages into fixed-length binary chains in a short and justified time. The hash algorithm finds its application in blockchain for system security, information authentication, PoW in consensus estimation, the interaction amongst the blocks, and several others. Blockchain stores complete metadata and can neither erase, remove, nor alter the blocks. The Merkle tree, mainly binary and multiform, is identical to the tree – as

TABLE 11.1
Comparison of Consensus Algorithms

Type of Consensus Algorithm	Algorithm Speed	Security	Energy Exploitation	Degree of Centralization
Proof of Work	Slow	Secure	Very high	Very low
Proof of Stake	Normal	Secure	Normal	Low
Delegated Proof of Stake	Normal	Secure	Normal	Normal
Practical Byzantine Fault Tolerance	Fast	Least secure	Very low	High
Proof of Elapsed Time	Normal	Secure	Low	Very low
Tendermint	Normal	Secure	Low	High

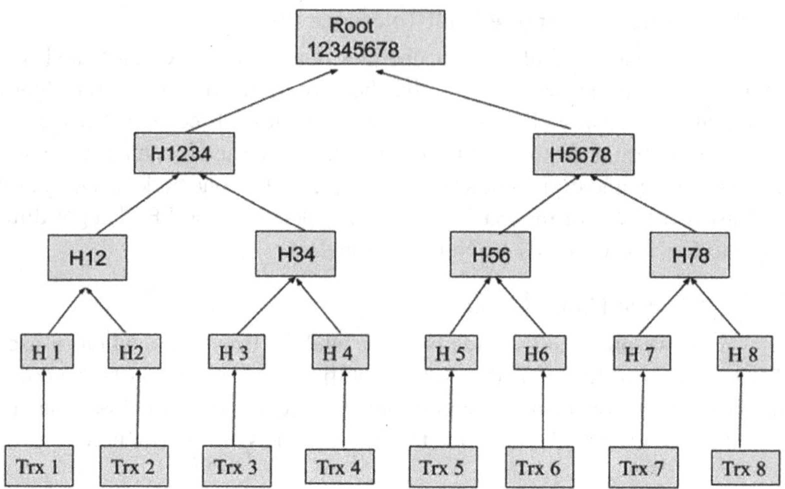

FIGURE 11.6 Merkle tree.

is the structure within the data structure. The last hash value is stored at Merkle's root in the stack (Wikipedia; hash function, 2020). Figure 11.6 explains the fundamentals of the Merkle tree.

11.2.4.8 Model Data Record

The blockchain recognizes two main frameworks for tracking a transaction's history: The UTXO model and the account model. UTXO is a process that underpins Bitcoin transaction development and authentication. Every unused output in the Bitcoin network is a UTXO. The model forecast is off-axis, with less line strain. However, such a complicated algorithm cannot be used in weak programming. In an account model, the account holds the balance. The chain's state is agreed as state root and receipt root in the block. The accountability model is programmable, and the costs of massive transactions are lower (Wilson and Ateniese, 2015; Dai et al., 2017).

11.3 BLOCKCHAIN APPLICATIONS IN THE PETROLEUM AND GAS SECTOR

Blockchain has applications in upstream, midstream and downstream sectors. Pipelines are considered lifelines that connect the three sectors (Maccallum, 2018; Arya and Shingan, 2012; Arya et al. 2015; Arya and Honwad, 2016, 2018a; Myalapilli et al., 2015; Thakur et al., 2020, 2021; Gupta et al., 2018, 2019a, b). According to Deloitte's reports, technology has an enormous opportunity in the petroleum and gas sector, particularly concerning four aspects: Trade, management decision, overvoice and cyber protection (Deloitte Blockchain Overview, 2017). Figure 11.7 shows the methodology adopted in implementing blockchain in the oil and gas sector. The following section discusses the potential implementation of these four elements.

TABLE 11.2
Major Pilot Programme Projects of Blockchain in the Oil and Gas Industry (S&P Global Platts. 2018)

Continent	Country	Company/Project	Details
Asia	Xiamen, China	Sinochem Group	Sinochem Energy Network Technology Co., Ltd., a joint venture of Sinochem Group, delivered a simulated export transaction for a ship carrying gasoline from China Quanzhou to Singapore in 2018. The project is the world's first blockchain-based oil trading project with government involvement. Additionally, it is the world's first blockchain project to include all stakeholders' interests in the commodity trading process.
	Abu Dhabi	ADNOC and IBM	ADNOC comprises 14 companies that trade and import goods using IBM's internal networking Hyperledger blockchain. Previously, accounting was performed manually; however, blockchain technology is now assisting the sector in achieving greater accounting transparency. Since it is a public ledger, anyone has access to accounting, and all changes made by everyone are available to everyone.
	Fujairah	FOIZ, S&P Global Platts	Fujairah Oil Industry Zone (FOIZ) has the highest industrial holding space for processed oil products in the Middle East. S&P Global Platts has established a fully commercial blockchain distributed ledger platform implementation. This provides FOIZ and its port authorities with security, ease of usage and a complete audit trail for collating weekly inventory data on oil product storage.
Europe	London	VAKT	VAKT is developing a distributed ecosystem for post-trade physical processing. Using blockchain technology would become a direct source of reality for the whole trading lifecycle by using blockchain technology. It would eradicate reconciliation and paper-based procedures, increase productivity and open up new markets for trade finance.
	Hamburg	Enerchain	The Enerchain project, led by multinational utility Enel, aims to establish a blockchain-based exchange that will allow energy wholesalers to list and sell anticipated future energy production.

(*Continued*)

TABLE 11.2 (CONTINUED)
Major Pilot Programme Projects of Blockchain in the Oil and Gas Industry (S&P Global Platts. 2018)

Continent	Country	Company/Project	Details
	Britain, Italy, Austria	Interbit	Interbit, as a next-generation blockchain network, is capable of operating and interconnecting thousands of blockchains on a single server in a stable, confidential and scalable manner.
	Switzerland	Komgo S.A.	Komgo was introduced in late 2018 to automate commodity market trading. Komgo is the world's first blockchain/distributed ledger technology (DLT)-based network for the financial trading of commodities.
North America	Houston	S&P Global Platts	S&P Global Platts, the largest independent supplier of knowledge and benchmark pricing for the commodities and energy markets, uses an encrypted blockchain network that allows market participants to send with increasing regularity oil storage data to FOIZ and the regulator, FEDCom.
South America	Chile	Energia Abierta	Energia Abierta is a blockchain project of the National Energy Commission that consists of a multi-functional Internet platform that has been built to address a broad spectrum of energy-related interests.

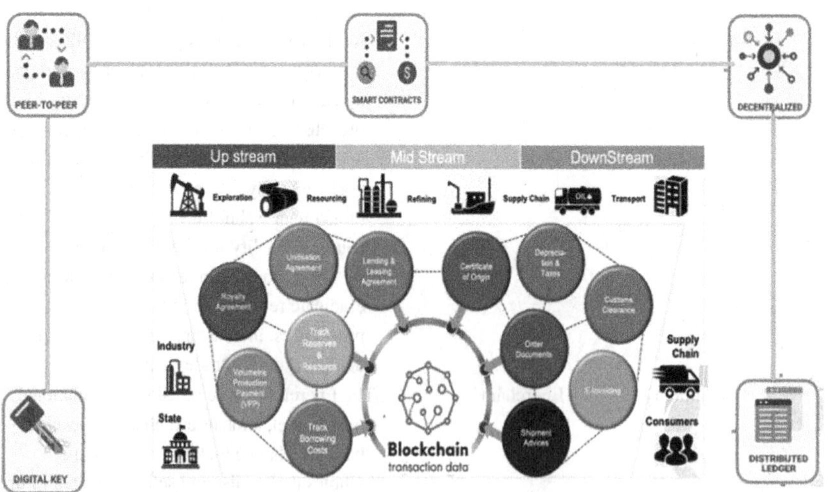

FIGURE 11.7 Implementing blockchain in petroleum and gas sector.

11.3.1 Trade

"International Oil and Gas Transactions Analysis 2017" reveals the petroleum and gas trade at 343.5 billion dollars (E.Y., 2017). The energy sector comprises a variety of related sub-industries. The vast volume of transactions and deals in these processes results in comprehensive arbitration work and surveillance work.

11.3.1.1 Smart/Intelligent Contract

Smart contracting is a form of acquiring. Smart contracts should include programming language rather than legal language. The oil and gas industry's complex nature is such that long and complicated contracts always exist. Intelligent contracts can significantly reduce red tape, automate operations, increase productivity and reduce costs. The smart contract should be verified before use and comply with Smart Contract Protection Standards, as poor architecture results in severe losses. In February 2018, numerous investigators in Singapore and the United Kingdom reported that there might be vulnerabilities to well over 34,000 intelligent contracts (Huatai Securities Oil and Gas Industry Chain Depth Report, 2018).

11.3.1.2 Transaction

In the oil and gas sector, the conventional method creates commerce errors and is vulnerable to corruption. The exchange may even be made simpler by using blockchain. Many processes are implemented, including opening the account and customs clearance. Money that is encrypted would significantly minimize the expense of international transfers, thereby reducing the time it takes intermediaries to authenticate and liquidate funds. Transaction intensity is often high and differs from the magnitude of inter-bank transactions. The purpose of trade may be covered, overlapping or interspecies arbitrage (Lakhanpal and Samuel, 2018; HIS Markit Oil and Gas Petcher Industries Blockchain solution, 2018).

11.3.1.3 Management Decisions

Blockchain also has a significant technological capacity to make decisions. Management decision-making requires multiple informed choices by the entire system's facts and evidence. Furthermore, specific petroleum and gas industries' decisions require voting management levels, and intelligent blockchain contracts allow automatic and open voting applications. Voters can vote or send their votes to others, and everyone can check the results publicly. The technology improves the reliability of data exchange and dissemination, leading to more accurate decisions. Data is recorded at each level that cannot be misused, contributing significantly to the design process (Blockchain empowers oil and gas industry, 2018). Pipeline transportation is an example of an intelligent management system. The pipeline system is complex and challenging to manage, especially concerning energy allocation. If relevant demand and supply details are generated and smart contracts signed, oil and gas deployment using pipelines can be scientifically reinforced.

11.3.1.4 Overvoice

11.3.1.4.1 Tracking of Petroleum Products

After the advent of liquefied natural gas (LNG) in the natural gas industry, its influence has rapidly taken a commanding position. Using dockets may decrease employee productivity and efficiency and trigger a substantial decrease in the amount of product generated for the business. Blockchain keeps track of petroleum and gas supply chains and monitors service facilities for all stages leading up to consumption. It addresses asset or property rights problems. However, due to the high legal cost of copyright ownership and the strong likelihood that copyright registrations will be later annulled, it is not cost-effective to register copyrights. It is also possible to track similar equipment using the same technique (Chritdis and Devetsikiotis, 2016).

11.3.1.4.2 Compliance

Blockchain innovations pave the door to boost oil and gas trading compliance. Technology is now addressing many of the challenges connected with the oil production and construction period. The blockchain data exchanged in the "Trust Data" system is trusted (Anjum et al., 2017).

11.3.1.4.3 Record of Data

Before the industry undertakes exploration, construction and other activities, petroleum and gas companies must obtain land use rights. Property transactions are highly fraudulent. Blockchain technology helps create a land mobility, value and ownership audit trail (Liang et al., 2017).

11.3.1.4.4 Cyber Security

Before undertaking exploration, production and other operations, petroleum and gas companies must acquire the right to use the property. The land deal is very likely to be fraudulent in this context. Blockchain technology creates a verification trail of mobility, value and property ownership. It eliminates the property's loss or absence and the impact of conflict on the property (Mittal et al., 2017).

11.4 STATUS OF BLOCKCHAIN AND PILOT PROGRAMMES

Many petroleum firms are already beginning to invest in developing this newest technology. BP and Shell are experts in keeping energy requirements at optimum efficiency. The Abu Dhabi National Oil Company (ADNOC) simplified the platform's purchasing process using advanced technology tools, and a saving of up to 30% was achieved in 2019. Sinochem was the first industry that executed the first pilot plant to import crude oil in China in 2017. The automated charge ratio and smart contract would significantly increase petroleum trade efficiency. VAKT Trading Crude Oil Consortium 2018 is a crude oil trading facility for petroleum giants like BP, Shell and Equinor. It replaces a manual paper system using fax, email or courier (BP and Shell Join Big Banks, 2019). In 2018, China-based VeChain launched a pilot programme for LNG quality control and operational management(Vechain Inks

Blockchain Deal for Chinese Gas Solutions, 2019). To trace, verify and carry out crude oil transactions, ADNOC is piloting a stable, blockchain-enabled platform for production wells to the customer. The pilot concentrates on ADNOC activities to simplify accounting procedures, minimize run-time and increase reliability (Adnoc Pilots Blockchain Program Across Value Chain with IBM, 2020).

11.5 LIMITATION AND CHALLENGES OF BLOCKCHAIN TECHNOLOGY

While blockchain technology has its advantages, the existing format is not flawless, and many risks arise. The four types of challenges that we can't ignore are threats to operations, cyber threats, regulatory hazards and legal risks. Risk managers expect that implementing blockchain can adversely affect the petroleum and gas sector. Blockchain is still fresh to the industry, with only minor failures as yet, although several other possible issues could plague the structure. Knowledge can be lost, and names are usually not accessible despite being requested. Public transfers can be quite large. Fraud exists between the public realm's truth and the blockchain's distributed ledger structures. There is a risk that EMV cards' potential to be exchanged for more money laundering should be checked to protect them. Unlawful contracts can be published into decentralized blockchains and impose tax evasion (Accenture Blockchain Benefit, 2018; Surujnath, 2017; Lindman et al., 2017; Walch, 2015; Cao., 2017). There are always several impediments to the innovation of new technology. Blockchain technology is also a newly adopted technology and faces challenges. This section identifies some of the significant challenges to applying blockchain in the petroleum and gas industry.

11.5.1 Evolution

With an increase in digital technology, the number of transactions is increasing. The blockchain gets more and more complex with every transaction. Already, the Bitcoin network has gone through 100 gigabytes. Just a single purchase may be authenticated and entered into the database. Bitcoin is a very tiny coin that cannot be traded for any of the value that we would want and cannot, by itself, fulfil billions of dollars of money at a time. If the market for Bitcoin approaches what is feasible to make, blockchain may become much less available for all who choose to use it for transactions. If the nodes are far away from each other, the blockchain is broken into several parts, and then it becomes impossible to resolve the transition. Several algorithms resolve these issues, one of which is scaling. To solve a complex problem, we need to optimize data via working parts while managing the outcomes. A novel scheme is implemented to resolve the security issue of storing data on the blockchain (Bruce, 2014). New transaction data is placed in a "trunk chain" on the network, enabling consumers to display their balance. A novel authentication system eliminates the need to keep all previous transactions in memory. The customer may even correct glitches in the website. The old client is replaced by a new system managed by the new scheme (Van de Hoof et al., 2014). "VerSum" is an algorithm for computing massive dense linear structures. It guarantees that the outcome is correct by comparing it against different servers. Eyal et al. (2016) proposed "Next-generation Bitcoin" for blockchain (Eyal et al., 2016). The purpose was to incorporate a component that does not influence another block key portion. A "strong block" is a node with a significant number of votes from the rest of the network. The block size is redesigned to retain both block size and transaction costs.

11.5.2 Data Theft

The blockchain is a very secure method to share digital tokens, since it only enables people to make transfers using a reference key. Users may create multiple addresses, making it difficult to trace when information leaks. However, Kosba et al. (2016) demonstrated that blockchains with a wallet, including a public key and transaction records, cannot guarantee transaction anonymity in an entirely general P2P model (Kosba et al., 2016). Besides, the study conducted by Barcelo (2014) suggested that it is possible to connect a user's Bitcoin transaction to disclose their details. Many approaches are proposed to enhance blockchains' privacy, divided into semi-anonymous and anonymous forms. On the blockchain, addresses record pseudonymous interactions. And with all of these safeguards, it is always possible to connect a speech to the consumer's actual identity via transaction history, so the user must be cautious. In semi-anonymous mixing, the transfer is made from many individuals to a single or many individuals without disclosing the person's identification (Ruffing et al., 2014). Miners are allowed to verify a transaction via digital signature. Instead of validating the coins themselves, the miner only makes sure that the blockchain's currency belongs to a valid coin registry. To avoid information leakage to the workstations, the payment's fund balance is not associated with the blockchain transaction. It still shows the sources and the sums of the expenses. One proposed method, named Zero cash, aims to overcome the Zero coin's privacy issue. In combination with zero-knowledge proofs, zero cash's lightning-speed transaction mechanism is brought to fruition. According to purchase quantities and user values, one does not realize how much a user is paying.

11.5.3 Greedy Exploitation

Blockchain is open to exploitation by greedy miners. As a general rule, it is thought that 51 percent of nodes could break the blockchain and that the transaction did in fact take place. However, the system is fragile even if it is not 51% controlled. The list can become unreliable if only a small amount of hash power is used. The miners excavate unrevealed blocks without exchanging the freshly found branches with any other groups. The private track is longer than the current public path, going into operation with fair approval from both miners. Miners' capital would be invested without private blockchains, although dishonest miners extract money without opponents. Selfish individuals gain more income than others. If a greedy miner has ownership of the pool, there is a fair risk that 51% mining will occur. There have been mining attacks that showed blockchain is not as reliable as has been supposed.

Sapirshtein et al. (2016) indicate more profitable greedy mining methods for smaller miners to contribute. They show that hackers can still obtain coins through fraudulent mining, even using remote computing capacity. Billah (2015) chose a new approach to motivate fair miners to pursue correct routes. The more truthful the miners, the more frequently they might pick new blocks. However, fabricated timestamps are a problem. Block creation and block adoption must occur within a short time frame, ensuring that the greedy miners won't get the benefit they anticipated.

11.6 POSSIBLE FUTURE DIRECTIONS

Blockchain has emerged as a potential technology to improve profitability in the oil and gas industry. Here, we present five primary considerations for the advancement of blockchain technology.

11.6.1 Testing Blockchain

Numerous new blockchains and cryptocurrencies have appeared recently throughout the news. However, developers with poor company practices could misrepresent their blockchain growth results to encourage investors. Besides, what counts is the user's desire: As companies incorporate blockchains, they must choose the more suitable blockchain. A verification framework has to be put into effect to validate various blockchain technologies. The distributed ledger can have two sub-stages: The standardizing and testing phases. The requirements and specifications are thought through and analyzed in the standardization process. When the blockchain and its protocol are updated, a Bitcoin developer group undertakes a peer review. We need to enhance the quality assurance standards in the engineering of blockchain systems.

11.6.2 Incentivize Decentralization

Blockchain is structured as an open, distributed and transparent ledger. The fact that more people are participating in mining pools may result in the best of all possible outcome. Currently, only five mining pools share 51% of the overall hash capacity (Szabo, 1997). Besides, the greedy mining strategy (Eyal and Sirer, 2014) reveals that networks with more than 25% of the total computing resources could receive more revenue than a fair portion of the network's strength. The selfish miners flood into the pit, and its share grows past 51% (Eyal and Sirer, 2014).

11.6.3 Large (Data-based) Analytics

Data processing comprises data production and data analysis. Blockchains can store critical data, as they are distributed and stable. The blockchain ensures that data is original. If the blockchain is used to keep records, it is unlikely to be modified. Blockchain transfers may be used for collecting data analytics. Users foresee potential trade partners.

11.6.4 Intelligent/Smart Contract

A smart contract is a hard-coded program that implements a predetermined contract's provisions. It was an idea generated through blockchain. The innovation of blockchain often included smart contracts that miners would run dynamically. This motive is the rationale behind the development of more intelligent contract sites. The IoT and finance are the targets of blockchain technology. Many sectors, such as banking, operate using blockchain technology (Peters et al., 2015, 2016). We expect intelligent contract analysis to fall into development and evaluation categories. The project could be an intelligent contract creation, implementing an intelligent contract framework, or both. In the Ethereum network, some intelligent contracts were introduced (Wood, 2014). Ethereum and Hawk have become popular platforms while developing an intelligent contract system. Evaluation applies to code interpretation and output evaluation. Bugs from intelligent contracts can cause catastrophic damage. As a result of recursive call errors, hackers stole over $60 million from the DAO (Jentzsch, 2016). DAO is 'Decentralized autonomous organization' formed on the Ethereum blockchain concepts. Thoughtful analysis of contract breaches is critical to performance. Intelligent contract efficiency is also essential for smart contracts. Blockchain technologies are now developing exponentially and are widely implemented in smart contracts. Companies must take the output of the application into account.

11.6.5 Artificial Intelligence

Current advancements in blockchain technologies provide exciting prospects for artificial intelligence (AI) ventures (Omohundro, 2014). AI technology could assist in solving several blockchain challenges. There is always an oracle tasked with deciding whether the contractual obligation complies. This seer is a reputable third source. Research on AI has the potential to help create a smarter predictor. Data is not regulated by any one institution, but rather is researched and educated externally.Thus, there are no disputes in the intelligent contract. AI is crucial now. Blockchain and smart contracts could discourage misuse and provide insight into AI benefits. Autonomous vehicle misbehaviour can be regulated using intelligent contracts.

11.7 CONCLUSION

This chapter extensively reviewed the potential uses of blockchain technology for the oil and gas industry. On the one hand, blockchain technology offers tremendous promise to the oil and gas sector, and on the other hand, it presents possibilities, obstacles and threats, as it has only begun in recent years. An energy blockchain may be a game-changer for industry and operations. The chapter emphasizes that blockchain use in the petroleum and gas sector is currently in the testing stage, and those in the petroleum and gas industry do not recognize it enough. Asia and Europe are the most prominent nations using blockchain in the petroleum and gas sector. BP and

Shell are leaders in this area. Blockchain initiatives can attract oil and gas companies because of their substantial variations. Blockchain technology implementation in the petroleum and gas industry will reduce transaction expenses and enhance accountability. It is expected that the network will shift into a more sophisticated blockchain and cross-connect with other blockchains, resulting in an enhanced hybrid consensus mechanism and even architecture to make it simpler to handle. Based on the fact that so many major players have already taken the lead in this sector, we can reasonably conclude that it's here to stay.

REFERENCES

Abe, R., Watanabe, H., Ohashi, S., Fujimura, S., & Nakadaira, A. (2018, July). Storage protocol for securing blockchain transparency. In 2018 IEEE 42nd annual computer software and applications conference (COMPSAC) (Vol. 2, pp. 577–581). IEEE.

Accenture. (2018). The blockchain benefit. Accenture, USA [Online]. Available: https://www.accenture.com/t20180521T235705Z__w__/us-en/_acnmedia/PDF-77/Accenture-13584-ACN-RES-Blockchain-FBAP-Brochure-2018-final.pdf.

Adnoc pilots blockchain program across the value chain with IBM. *The National.* [Online]. Available: https://www.thenational.ae/business/energy/adnoc-pilots-blockchain-programme-across-value-chain-with-ibm-1.800755

Anjum, A., Sporny, M., & Sill, A. (2017). Blockchain standards for compliance and trust. *IEEE Cloud Computing*, *4*(4), 84–90.

Arya, A.K., & Honwad, S. (2016). Modeling, simulation, and optimization of a high-pressure cross-country natural gas pipeline: application of an ant colony optimization technique. *Journal of Pipeline Systems Engineering and Practice*, *7*(1), 04015008.

Arya, A.K., & Honwad, S. (2018a). Multiobjective optimization of a gas pipeline network: an ant colony approach. *Journal of Petroleum Exploration and Production Technology*, *8*(4), 1389–1400.

Arya, A.K., & Honwad, S. (2018b). Optimal operation of a multi-source multi-delivery natural gas transmission pipeline network. *Chemical Product and Process Modeling*, *13*(3), 20170046.

Arya, A.K., & Shingan, B. (2012). Scour-mechanism detection and mitigation for subsea pipeline integrity. *International Journal of Engineering Research & Technology*, *1*(3), 1–14.

Arya, A.K., B. Shingan, & C. Vara Prasad. (2015) Seismic design of the continuous buried pipeline. *International Journal of Engineering Science*, 1 (1): 6–17.

Atlam, H.F., & Wills, G.B. (2019). Technical aspects of blockchain and IoT. In *Advances in Computers* (Vol. 115, pp. 1–39). Elsevier.

B.P. and Shell Join Big Banks for Blockchain Project 1 VAKT [Online]. Available: https://www.vakt.com/oil-industry-backed-blockchain-energy-platform-launching-this-month/ [Accessed: 15-Oct-2019].

B.P. Energy Outlook. (2019). Retrieved from https://www.bp.com/content/dam/bp/business-sites/en/global/corporate/pdfs/energy-economics/energy-outlook/bp-energy-outlook-2019.pdf

Barcelo, J. (2014). User privacy in the public bitcoin blockchain. URL: http://www. dtic. upf. edu/jbarcelo/papers/20140704 User Privacy in the Public Bitcoin Blockchain/paper.pdf (Accessed 09/05/2016).

Billah, S. (2015). *One weird trick to stop selfish miners: Fresh bitcoins, a solution for the honest miner.*

Blockchain. (2018). Blockchain empowers the oil and gas industry or will boost new energy alternatives to traditional energy sources. [Online]. Available: https://mp.weixin.qq.com/s?__biz=MzU5OTQ1MTM4NA==&mid=2247484250&idx=1&sn=2ce5c0c2314ee84594fdc9b06e7703db&chksm=feb5f0cdc9c279db1babc5088fff01467b768ec4267805d86fc5f38d68200eaf5fd15db97006&mpshare=1&scene=23&srcid=0108Vj31D6Gyv4dRLLgH0EgK#rd.

Bruce, J. (2014) *The mini-blockchain scheme.* http://cryptonite.info/files/mbc-scheme-rev3.pdf

Cao, X. Be wary of the legal risks contained in the blockchain. 19-Dec.-2017. [Online]. Available: https://mp.weixin.qq.com/s?__biz=MjM5MDIxODEzOQ==&mid=2650185097&idx=2&sn=e69a4bc1e392e8fbb9bb1ec032f866f6&chksm=be4a1f7d893d966b7ccdc56237a9da0ff8dfad05a93caaef7febabb9e1c01b491ae173f793ac&mpshare=1&scene=23&srcid=01154XvkDdivoW3BeFJpX5us#rd.

Casino, F., Dasaklis, T.K., & Patsakis, C. (2019). A systematic literature review of blockchain-based applications: current status, classification, and open issues. *Telematics and informatics, 36*, 55–81.

Christidis, K., & Devetsikiotis, M. (2016). Blockchains and smart contracts for the internet of things. *IEEE Access, 4*, 2292–2303.

Dai, P., Mahi, N., Earls, J., & Norta, A. (2017). *Smart-contract value-transfer protocols on a distributed mobile application platform.*

Deloitte. (2017). Blockchain: Overview of the potential applications for the oil and gas market and the related taxation implications. [Online]. Available: https://www2.deloitte.com/content/dam/Deloitte/global/Documents/Energy-and-Resources/gx-oil-gas-blockchain-article.pdf

Deloitte. (2018). Deloitte's 2018 global blockchain survey. Deloitte, USA. [Online]. Available: https://www2.deloitte.com/content/dam/Deloitte/cz/Documents/financial-services/cz-2018-deloitte-global-blockchain-survey.pdf.

Dorri, A., Roulin, C., Jurdak, R., & Kanhere, S.S. (2019, October). On the activity privacy of blockchain for IoT. In 2019 IEEE 44th Conference on Local Computer Networks (LCN) (pp. 258–261). IEEE.

Dütsch, G., & Steinecke, N. (2017). Use cases for blockchain technology in energy and commodity trading. *Snapshot of current developments of blockchain in the energy and commodity sector. pwc.*

E.Y. (2018). Global oil and gas transactions review 2017. E.Y. [Online]. Available: https://www.ey.com/Publication/vwLUAssets/ey-global-oil-and-gas-transactions-review-2017/$FILE/ey-global-oil-and-gas-transactions-review-2017.pdf.

Eyal, I., & Sirer, E.G. (2014, March). Majority is not enough: Bitcoin mining is vulnerable. In International Conference on Financial Cryptography and Data Security (pp. 436–454). Berlin, Heidelberg: Springer.

Eyal, I., Gencer, A.E., Sirer, E.G., & Van Renesse, R. (2016). Bitcoin-ng: A scalable blockchain protocol. In 13th {USENIX} symposium on networked systems design and implementation ({NSDI} 16) (pp. 45–59).

Fraser, M.S., Anastaselos, T., & Ravikumar, R. (2018). *The disruption in oil and gas upstream business by industry 4.0.* Bengaluru, India: Infosys, White Paper.

Gupta, S.S., Arya, A.K., Vijay, P., & Kumar, S. (2019a). Control of microbial induced corrosion in an oilpipeline: A case study. *Journal of the Gujarat Research Society, 21*(2), 355–361.

Gupta, S.S., Arya, A.K., Vijay, P., & Kumar, S. (2019b). Designing a model for optimization of maintenance and inspection efforts against third party damage to cross country pipelines in India. *International Journal of Innovative Technology and Exploring Engineering, 8*(12), 4529–3539.

Hamida, E.B., Brousmiche, K.L., Levard, H., & Thea, E. (2017). Blockchain for enterprise: overview, opportunities and challenges. In The Thirteenth International Conference on Wireless and Mobile Communications (ICWMC 2017).

HIS Markit. (2018). Oil and gas, petchem industries trialing blockchain solutions. HIS Markit, United Kingdom. [Online]. Available: https://cdn.ihs.com/www/pdf/IHS-Markit-Blockchain%20trials%20in%20OG-Petchem.pdf.

Huatai Securities. (2018). Oil and gas industry chain depth report. Huatai Securities. [Online]. Available: https://crm.htsc.com.cn/doc/2018/10720102/9cca9c98-df91-456d-90bf-6a2cdc656666.pdf.

Hughes, E., Graham, L., Rowley, L., & Lowe, R. (2018). Unlocking blockchain: Embracing new technologies to drive efficiency and empower the citizen. *The Journal of The British Blockchain Association*, *1*(2), 3741.

Jansen, A. (2018). *Investigation of the potential of blockchain technology to create traceability and transparency along a raw material value chain.*

Jentzsch, C. (2016). The history of the DAO and lessons learned. *Slock. it Blog*, *24*.

Kosba, A., Miller, A., Shi, E., Wen, Z., & Papamanthou, C. (2016, May). Hawk: The blockchain model of cryptography and privacy-preserving smart contracts. In 2016 IEEE symposium on security and privacy (S.P.) (pp. 839–858). IEEE.

Lakhanpal, V., & Samuel, R. (2018, September). Implementing blockchain technology in oil and gas industry: A review. In SPE Annual Technical Conference and Exhibition. Society of Petroleum Engineers.

Liang, X., Shetty, S., Tosh, D., Kamhoua, C., Kwiat, K., & Njilla, L. (2017, May). Provchain: A blockchain-based data provenance architecture in cloud environment with enhanced privacy and availability. In 2017 17th IEEE/ACM International Symposium on Cluster, Cloud and Grid Computing (CCGRID) (pp. 468–477). IEEE.

Lindman, J., Tuunainen, V.K., & Rossi, M. (2017). *Opportunities and risks of Blockchain Technologies: a research agenda.*

MacCallum, T. (1-Feb. 2018). Diving into Ethereum's world state. [Online]. Available: https://medium.com/cybermiles/diving-into-ethereums-world-state-c893102030ed.

Midstream companies and the oil & gas downturn deloitte insights. [Online]. Available: https://www2.deloitte.com/us/en/insights/industry/oil-and-gas/decoding-oil-gas-downturn/midstream-pipeline-infrastructure-transportation.html#endnote-sup-3.

Mittal, A., Slaughter, A., & Zonneveld,P. (2017). "Protecting the connected barrels: Cybersecurity for upstream oil and gas," Deloitte Insights, London, U.K., Tech. Rep., 2017. Available: https://www2.deloitte. com/insights/us/en/industry/oil-and-gas/cyber security-in-oil-andgas-upstream-sector.htm

Mylapilli, L.K., Gogula, P.V.R., & Arya, A.K. (2015). Hydraulic and surge analysis in a pipeline network using pipeline studio®. *International Journal of Engineering Research and Technology*, *4*(2), 382–389.

Omohundro, S. (2014). Cryptocurrencies, smart contracts, and artificial intelligence. *AI Matters*, *1*(2), 19–21.

Peters, G., Panayi, E., & Chapelle, A. (2015). Trends in cryptocurrencies and blockchain technologies: A monetary theory and regulation perspective. *Journal of Financial Perspectives*, *3*(3), 99–113

Peters, G.W., & Panayi, E. (2016). Understanding modern banking ledgers through blockchain technologies: Future of transaction processing and smart contracts on the internet of money. In *Banking beyond banks and money* (pp. 239–278). Cham: Springer.

Pilkington, M. (2016). Blockchain technology: principles and applications. In *Research handbook on digital transformations*, edited by F. Xavier Olleros and Majlinda Zhegu. Available at SSRN 2662660. https://masterthecrypto.com/public-vs-private-blockchain-whats-the-difference/

Puthal, D., Malik, N., Mohanty, S.P., Kougianos, E., & Yang, C. (2018). The blockchain as a decentralized security framework [future directions]. *IEEE Consumer Electronics Magazine*, *7*(2), 18–21.

Ruffing, T., Moreno-Sanchez, P., & Kate, A. (2014, September). Coinshuffle: Practical decentralized coin mixing for bitcoin. In European Symposium on Research in Computer Security (pp. 345–364). Cham: Springer.

S&P Global Platts. (2018). Blockchain for commodities: Trading opportunities in a digital age. United Kingdom [Online]. Available: https://www.spglobal.com/platts/en/market-insights/specialreports/oil/blockchain-for-commodities-trading-opportunities-in-adigital-age.

Samaniego, M., Jamsrandorj, U., & Deters, R. (2016, December). Blockchain as a Service for IoT. In 2016 IEEE international conference on internet of things (iThings) and IEEE green computing and communications (GreenCom) and IEEE cyber, physical and social computing (CPSCom) and IEEE smart data (SmartData) (pp. 433–436). IEEE.

Sapirshtein, A., Sompolinsky, Y., & Zohar, A. (2016, February). Optimal selfish mining strategies in bitcoin. In International Conference on Financial Cryptography and Data Security (pp. 515–532). Berlin, Heidelberg: Springer.

Sarmah, S.S. (2018). Understanding blockchain technology. *Computer Science and Engineering*, *8*(2), 23–29.

Sikorski, J.J., Haughton, J., & Kraft, M. (2017). Blockchain technology in the chemical industry: Machine-to-machine electricity market. *Applied Energy*, *195*, 234–246. doi:10.1016/j.apenergy.2017.03.039

Sultan, K., Ruhi, U., & Lakhani, R. (2018). Conceptualizing blockchains: characteristics & applications. arXiv:1806.03693.

Surujnath, R. (2017). Off the chain: A guide to blockchain derivatives markets and the implications on systemic risk. *Fordham Journal of Corporate & Financial Law*, *22*, 257.

Szabo, N. (1997). The idea of smart contracts. *Nick Szabo's Papers and Concise Tutorials*, *6*(1). *http://www.fon hum uva.nl/rob/Courses/InformationInSpeech/CDROM/Literature/LOTwinterschool2006/szabo. best. vwh. net/smart contracts* 2.

Thakur, A.K., Arya, A.K., & Sharma, P. (2020). The science of alternating current-induced corrosion: a review of literature on pipeline corrosion-induced due to high-voltage alternating current transmission pipelines. *Corrosion Reviews*, *38*(6), 463–472.

Thakur, A.K., Arya, A.K., & Sharma, P. (2021). Analysis of cathodically protected steel pipeline corrosion under the influence of alternating current. *Materials Today: Proceedings*. DOI:10.1016/j.matpr.2021.05.548

Uba, S. (2021). *Blockchain adoption in the oil and gas supply chain.*

van den Hooff, J., Kaashoek, M.F., & Zeldovich, N. (2014, November). Versum: Verifiable computations over large public logs. In Proceedings of the 2014 ACM SIGSAC Conference on Computer and Communications Security (pp. 1304–1316).

VeChain inks blockchain deal for Chinese Gas Solution: Ledger Insights. [Online]. Available: https://www.ledgerinsights.com/vechain-blockchain-china-liquefied-natural-gas/. [Accessed:15-Oct-2019].

Walch, A. (2015). The bitcoin blockchain as financial market infrastructure: A consideration of operational risk. *NYUJ Legislation & Public Policy*, *18*, 837.

Wikipedia. Hash function. [Online]. Available: https://en.wikipedia.org/wiki/Hash_function.

Wilson, D., & Ateniese, G. (2015, November). From pretty good to great: Enhancing PGP using bitcoin and the blockchain. In International Conference on Network and System Security (pp. 368–375). Springer, Cham.

Wood, G. (2014) *Ethereum: A Secure Decentralized Generalized Transaction Ledger, Ethereum Project Yellow Paper.*

World Economic Forum, and Accenture. (2017). Digital transformation initiative oil and gas industry. World Economic Forum, Switzerland. [Online]. Available: http://reports.weforum.org/digital-transformation/wp-content/blogs.dir/94/mp/files/pages/files/dti-oil-and-gas-industry-white-paper.pdf

Yuan, Y., & Wang, F.Y. (2016). Blockchain: the state of the art and future trends. *Acta Automatica Sinica*, *42*(4), 481–494.

Zambrano, B. (2018) Blockchain explained: How does immutability work?. Available: https://www.verypossible.com/blog/blockchain-explained-how-does-immutability-work.

Zheng, Z., Xie, S., Dai, H., Chen, X., & Wang, H. (2017). An overview of blockchain technology: Architecture, consensus, and future trends. In 2017 IEEE international congress on big data (BigData congress) (pp. 557–564). IEEE, Honolulu.

12 Blockchain Technology in the Banking System in Developing Countries

Potential and Future Trends

Van Chien Nguyen, Lam Oanh Ha and Sarfraz Hussain

CONTENTS

12.1 INTRODUCTION

The Industrial Revolution 4.0 was first listed in Germany in 2011 as a proposal to implement a new paradigm focused on high-tech strategies in economic policy (Mosconi, 2015). Indeed, interoperability practices are the basis for the launch of the fourth technological revolution, focused on principles and technologies like physical network networks, the Internet of Things (IoT) and the Internet of Services (IoS). The Internet link enables constant connectivity and information sharing between individuals and blockchain technology in socio-economic activities (Mehdiabadi et al., 2020).

The revolution is changing the way of life today, but it also significantly affects the financial, banking and global industries. The Industrial Revolution 4.0 focuses mainly on using large-scale devices and deploying advanced automation, networking and surveillance across the Internet. In the World Economic Forum report (2016), $1 invested in digital technology has increased gross domestic product (GDP) by $20

DOI: 10.1201/9781003138082-12

over the past 30 years, while the same amount spent on non-digital investment has increased GDP by just $3. By 2025, 24.3% of global GDP is projected to come from digital technologies such as artificial intelligence and cloud computing.

Nevertheless, the Industrial Revolution 4.0 can increase global income and boost people's living conditions worldwide. Nowadays, people can afford to pay to access the digital world. It has helped improve productivity with new technology such as the IoT, artificial intelligence, cloud computing, big data, etc. Everybody's everyday life benefits. This technical breakthrough would bring greater access to the digital world over the next few years, with long-term benefits in quality and productivity. Transportation and communication costs are expected to decrease. Global logistics and supply chains are expected to become more productive, and trade is expected to minimize costs, open up new business opportunities and foster growth. Growth in economic terms. In creating Industry 4.0, Information and Communication Technology (ICT) forms the basis of the infrastructure for the future of innovative industrial technologies, including industrial integration and information integration. News plays an essential role in this regard (MacDougall, 2014; Gupta & Vyas, 2021).

Blockchain is a technology that enables data to be transferred safely based on an incredibly complex encryption scheme, similar to the ledger of a business, where the money is closely monitored. Blockchain is an accounting ledger running in the digital domain in this case (Osmani et al., 2020; Cermeño, 2016; Higginson et al., 2019). The specific feature of blockchain is that the transmission of data does not need an intermediary to validate the details. The most practical benefit of blockchain technology is that it provides us with a secure, fast and low-cost solution (Drescher, 2017; Osmani et al., 2020; Mendling et al., 2018; Cermeño, 2016; Mainelli and Smith, 2015).

Over half of today's top managers agree that blockchain plays a vital role in the performance of banks and financial firms, according to a survey by Accenture (2017). Analysts have stressed that by implementing blockchain technology, banks worldwide could save $20 billion by 2022. Some financial experts think that blockchain will soon replace current systems for bank transfer. A customer identity system based on a distributed ledger is another application that blockchain brings to banks. In a simple step, blockchain enables users to verify their identity, and that information is stored and approved in the system by other banks (Vovchenko et al., 2017).

12.2 BLOCKCHAIN TECHNOLOGY: A HISTORICAL OVERVIEW

The world has made great strides in all respects through the four technological revolutions shown in Figure 12.1. With the aid of steam energy, mechanical manufacturing facilities were built in the First Industrial Revolution. Mass manufacturing was achieved with the help of electrical power during the Second Industrial Revolution. The advent of electronic and information technology spurred the automation of production during the Third Industrial Revolution. The Fourth Industrial Revolution's development pattern is that physical channel systems (CPS) have dramatically changed production models in industries, especially in the manufacturing sector.

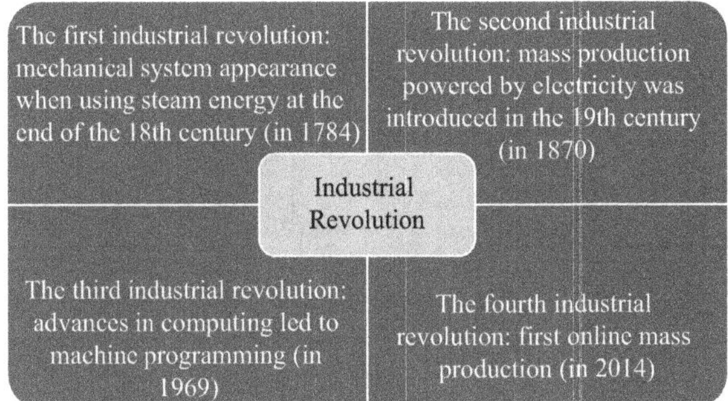

The first industrial revolution: mechanical system appearance when using steam energy at the end of the 18th century (in 1784)
The second industrial revolution: mass production powered by electricity was introduced in the 19th century (in 1870)
Industrial Revolution
The third industrial revolution: advances in computing led to machine programming (in 1969)
The fourth industrial revolution: first online mass production (in 2014)

FIGURE 12.1 History of industrial revolutions. (Synthesis by authors.)

Cherukupally (2020) argues that in the real world, peer-to-peer transactions can occur without the notice and intervention of a trusted third party between financial sources, especially money between two individuals. With the growth of the Industrial Revolution 4.0, the rapidly evolving environment is closely connected to the digital world. A responsible third party, such as a bank, still manages money between two online business entities and financial institutions. As suggested in the Bitcoin cryptocurrency strategy by Nakamoto (2009), which was seen as a turning point that began to shape the notion that peer-to-peer transactions could be implemented into the digital world, Bitcoin has become increasingly popular but not yet fully embraced as a means of liquidity.

Many previous studies have analyzed the benefits of blockchain, and they can be categorized into three key advantages, as suggested in Figure 12.2: (i) Decentralization of the process when money can be exchanged instantly between the parties without going through a bank, (ii) it cannot be changed or manipulated because the distributed ledger can be spread around a computer network, and (iii) transparency; everyone can see all transactions. According to Nakamoto (2009), blockchain technology would bring several benefits, such as increased transparency of data, increased transaction accountability and ease of control, high reliability level, etc. Transactions are transactions without intermediaries, so transactions are swift, and transaction costs are minimized, allowing the financial and banking industries to make the most of this new technology.

Garg et al.'s (2020) research assesses the potential market advantages of the banking and finance industries' adoption of blockchain technology. In particular, data is collected on blockchain consultants, blockchain marketing experts, CEOs, business directors working in the banking and finance sectors, security assessment, value systems and objectives. For banking, standards are essential. The research results confirm that blockchain technology's recent developments could impact research outcomes when technology such as blockchain is in its early stages. Besides, Garg et al. (2020) also say that developed instruments will provide decision-makers with a basic understanding of how to assess the advantages of incorporating blockchain technology before deciding to incorporate it into their current systems.

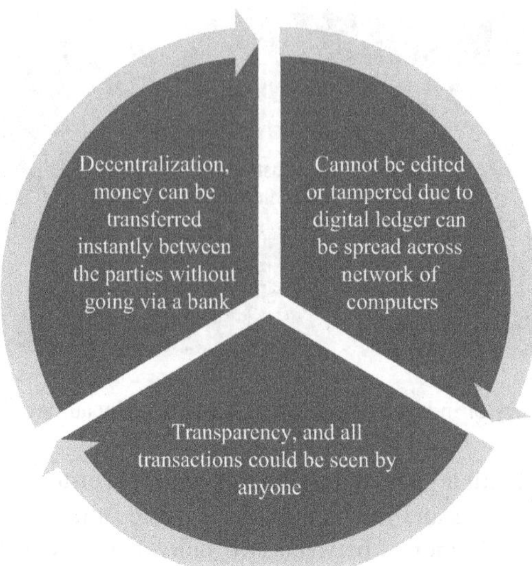

FIGURE 12.2 Three pillars of blockchain. (From Nakamoto, S., Bitcoin: A peer-to-peer electronic cash system, 2009, available at https://bitcoin.org/bitcoin.pdf; Cherukupally, S., (2020). *Handbook of Statistics*, 2020, available at https://doi.org/10.1016/bs.host.2020.10.001)

In another study by Chang et al. (2020), blockchain is a common trend today in the financial sector. Via in-depth interviews with 16 experts in the banking and finance sector, the author adopts the qualitative approach, is informed about blockchain and stresses that the financial industry is using a framework on the blockchain platform in the modern economic age. Previously, the goods and services suggested by the financial sector were deemed expensive and inefficient, imposing a significant change in the use of technology. Also, on a platform that creates confidence among transaction entities, blockchain can encourage credit reconstruction, a consensus mechanism that enables transactions to be correct. Blockchain technology also leads to enhancing financial markets' effectiveness and stability. In their growth, appropriate policies should be established by the government and related departments to allow blockchain to effectively bring functional benefits and prevent the illicit use of blockchain in currency laundering activities and terrorism support. Chang et al. (2020) also argue through in-depth interviews that knowledge concealment is a severe problem that can hinder the growth and progress of the adoption of blockchain, since hiding knowledge is due to employees' fear of losing control or that it may impact their current roles or jobs. It is also difficult for top management to consider new ideas, which is also the case with blockchain.

Blockchain architecture has many advantages not only in developed countries but also in developing countries. Tyson and Lund (2016) argued that globalization is entering a new age that is improving investment in low-income countries, not just influenced by the movement of goods and resources across borders. To encourage multinational companies to manufacture at a low cost, take advantage of globalization. From the World Bank's estimate (2020), global merchandise exports have contributed into 38% global GDP in 1980 to 38.8%, 51% in 1990, and 60.3% in 2010. From 2014 onwards, product movements, services, finance, citizens and cross-border data have increased global GDP by approximately 10%, including significant contributions from developed and emerging economies. As shown by the University of Pennsylvania (2018), the global economy's new trends are developing and emerging economies, such as India and Africa (Kenya, East Africa), where they are discovering more and more blockchain applications. Kshetri and Voas (2018) assess that blockchain can help a large proportion of the population in developed countries, so there is a need to concentrate more on critical issues in driving blockchain growth. Organizations worldwide are evolving and understanding the potential role of blockchain to help implement these advantages in today's lives.

12.3 BLOCKCHAIN TECHNOLOGY AND ITS POTENTIAL IN THE BANKING SYSTEM

Many recent studies indicate that, not only in developed countries but also in emerging and developing countries, there are many advantages of blockchain in the banking industry. According to a survey by Accenture (2017), blockchain plays a crucial role in the growth of both banks and financial firms. Banks worldwide could save 20 billion dollars by 2022 by implementing a banking system that facilitates faster payments and transfers and more accurate transactions using distributed ledger-based customer identification systems. With only a straightforward step, blockchain will enable users to validate their identity, and this information is stored in the system.

Blockchain is taking advantage of distributed ledger technology, according to Belinky et al. (2015), because it offers many benefits. Some of the advantages that blockchain technology brings to the banking industry are seen in Figure 12.3, with six key advantages, including irrevocable banking transactions, prompt payments, easy payments and transfers in the banking sector, as well as low-risk transactions. Also, transactions are often low cost, allowing the financial system and clients to save

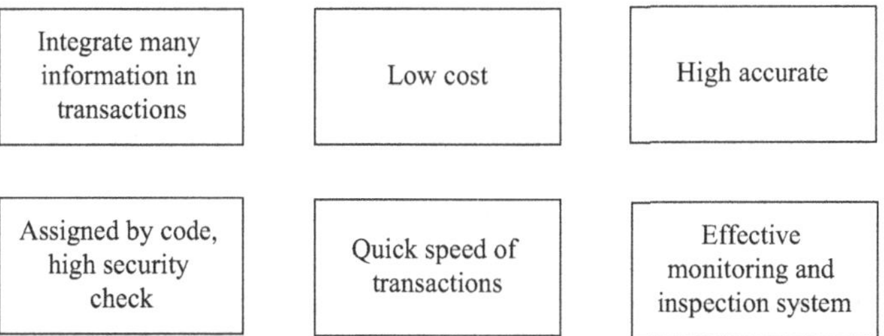

FIGURE 12.3 Benefits of using blockchain in transactions in the banking system. (From Belinky, M., et al., The fintech 2.0 paper: Rebooting financial services, 2015, retrieved from http://santanderinnoventures.com/wp-content/uploads/2015/06/The-Fintech-2-0-Paper.pdf; Kshetri, N., and Voas, J., *IEEE*, 20, 11, 2018.)

costs. The monitoring system via blockchain is even better, since each transaction is linked with high-security codes and the ledger system, with each transaction publicly checked by a network user group and the distributed ledger being tamperproof.

In recent years, some countries' regulators and policymakers have begun to attach importance to developed and developing countries' blockchain technology to keep up with the technology and close the development gap. Developed countries are also at the forefront of financial technology, such as the platform's recent substantial investment in integrating transactions for blockchain systems in Europe, North America and East Asia. Gradually, this technology is being applied. Banco Santander's Spanish banking group has applied blockchain technology to its One Pay F.X. (Santander, 2019) payment system. Morgan Stanley (2018) has invested in blockchain for the United States to catch the potential trend by setting up a company to study and apply this technology. Similarly, for the seven largest European banks (Deutsche Bank, HSBC, KBC, Natixis, Rabobank, Societe Generale and Unicredit), IBM is developing dedicated blockchain technology for efficient foreign trade. For the Asian market, China is leading blockchain research and application; as early as 2017, 12 out of 26 Chinese banks confirmed that they were using blockchain applications (People's Bank of China, 2017).

Seizing the benefits of blockchain transactions, primarily through the spillover of foreign bank productivity that will take place in developed countries, will bring foreign banks with modern technology to developing countries, creating interaction with domestic banks. IBM cooperated with the Thai government to implement blockchain technology in government bond management. Some banks such as Bangkok Bank, Krungthai Bank, Kasikorn Bank and Siam Commercial Bank (The Bank of Thailand, 2020). Similarly, Krungsri (Bank of Ayudhya Pcl.) became the first bank in Thailand to offer blockchain-based real-time international money transfers, and this is considered an essential platform for the bank to focus on future development.

India is also making many efforts to make its economy more transparent and reform-oriented, with Indian banks relying on blockchain technology to carry out transactions (Dhar and Bose, 2016). Also, overseas transactions, remittance transfers and other virtual currency payment procedures have been performed at ICICI Bank, Access Bank and Yes Bank. Most importantly, blockchain has been at the forefront of technology and growth activities in the Indian market and has also participated in the Morgan Stanley Blockchain Network (Dhar and Bose, 2016). Additionally, domestic banks have also strengthened cooperation to take advantage of the rapidly growing Indian economy, with 11 significant banks setting up a joint venture to cooperate and operate a blockchain lending system for small and medium-sized enterprises (SMEs). In contrast, the State Bank of India (SBI) has the support to establish these activities (Manikandan, 2019).

To accept the technology trend in developing countries in Africa, banks here have continuously promoted innovation and mobile payments in some areas and have the potential for innovative technology. The everyday lives of Africans have significantly changed. The World Bank (2017) estimates that many blockchain startups in Africa have adapted to this trend, such as Kenya's Bitcoin and Bitcoin, Ghana's Bitcoin Exchange, Lono and Ice3X platforms, and GeoPay, BitSur and Chankura, which connect Bitcoin payments and national stores, especially in South Africa. According to McKenzie Baker (2018), the number of mobile phone users has risen from less than 3% to 80% within a decade, with African countries increasingly becoming familiar with digital solutions to move money. However, there are still 350 million people without a bank account in some countries, such as Sub-Saharan Africa, which has not yet caught up with technology development, representing 17% of the global total. Therefore, the advantages of blockchain in the region are very different, with a newly formed financial market for countries such as Zimbabwe and Namibia. At the same time, Mauritius is a leader in using blockchain technology in the region.

Similarly, Kenya is also evolving in payment and banking operations to become a place to apply several blockchain technologies. Ghana is considering using blockchain to register land, and several nations want to exploit blockchain in digital identity, supply chain routes, healthcare and finance. The technology platform of FlexID aims to facilitate a blockchain-based digital identity running on the Algorand protocol and cryptocurrency transactions for Zimbabwe as the critical driver of the country's blockchain technology.

12.4 FUTURE TRENDS IN USING BLOCKCHAIN TECHNOLOGY IN THE BANKING SYSTEM

The implementation of banking and finance technology has enabled business transactions in various sectors (Peters and Panayi, 2016; Frame and White, 2014) and generated innovations in commerce and business (Ceremeno, 2016). In recent times, studies by Nakamoto (2009), Cherukupally (2020), Garg et al. (2020) and Chang et al. (2020) and the blockchain transition trend confirm that as the banking industry operates on the blockchain payment network, transactions will continue to expand, and this is a prominent trend in the financial sector. The birth of blockchain emphasizes that the financial industry is in a highly evolved era of fintech development, marking the world economy's transition and transformation. It is regarded as one of society's future technologies, especially as the Internet and technology 4.0 are increasingly emerging in a strong globalization trend. Indeed, individuals will be exposed to a massive, highly protected data system with blockchain technology advancement, where anything can become a service. Blockchain technology is intended to extend partnerships, databases and social interactions further and, in particular, to save costs because transactions are directly exchanged between two parties, not by any third party.

Blockchain technology has resulted in several financial and banking shifts in developed and developing countries, bringing many benefits. This trend will continue to evolve in the future. In Zimbabwe and Namibia, implementing a modern blockchain network was in its infancy when the financial industry began to expand. Still, the emergence of technology firms created opportunities for these nations. Vietnam is also starting to experiment with blockchain in banking and influence the supply chain in Asia. HSBC Bank successfully launched letter credit transaction (L/C) on the blockchain platform. The trade on the blockchain platform between Duy Tan Plastic Manufacturing Joint Stock Company (Vietnam) and INEOS Styrolution

Korea (Korea) saw improved operating efficiency and higher levels of transparency and protection for both companies (HSBC, 2020). Blockchain is developed on a distributed peer-to-peer system. However, the implementation of blockchain and its application in banking also face some challenges, particularly as anyone in the network can read transactions and add data. It has a negative database relationship and restricts the use of blockchain, impacting the security of data and transactions of customers in the banking system (Drescher, 2017). In addition, blockchain can generate several addresses to prevent information leakage instead of real identities for users. Still, blockchain has not managed transaction information leakage sufficiently to ensure customer protection; in addition to using blockchain applications, computer systems with extensive data and powerful configurations are still needed. These problems are not solved in the best way (Chang et al., 2020). The evaluation of blockchain' contribution to the banking and finance industries is often more constrained than in other fields, such as Osmani et al. (2020).

Although the blockchain platform is built on distributed ledger technology, it is a future advancement that is revolutionary and potentially game-changing. Nevertheless, according to the World Bank (2019), banking services in emerging and developing countries are restricted, business infrastructure is developed, and the demand for services in financial services is increasing rapidly, leading to blockchain. It will lead to increased employment in the financial technology field and increased interdisciplinary convergence between ICT, finance and banking. Many firms in the banking sector would be replaced immediately, resulting in the disappearance of specific financial and banking industries and leading to the loss of many financial and banking staff. Hanushek (2013) considers that focusing on creating human capital can be a driving force for developed countries' economic growth and that it is easier for workers to respond faster to changes. In the field of technology and careers, by investing in human capital by improving the quality of schools, developing countries have narrowed the gap in academic achievement with developed countries and stimulated economic growth. However, improving the quality of human capital in developing countries is becoming increasingly complex and linked to each country's economic growth. In the Human Development Index (HDI) from 1990 to 2019, Sub-Saharan Africa, South Asia and most other developing countries also are not keeping up with the world average, according to UNDP (2020). Excluding Singapore, the world's average HDI is 0.728, higher than the average HDI of developed countries (0.681). The five countries with the highest HDI are Norway (0.953), Switzerland (0.944), Australia (0.939), Ireland (0.938) and Germany (0.936). In comparison, the five countries with the lowest HDI are Burundi (0.417), Chad (0.404), South Sudan (0.388), the Central African Republic (0.367) and Niger (0.354).

While several reports have already confirmed the development of blockchain in the banking industry in developing countries, it continues to increase. But, legal issues are arising in some nations. First of all, the banking system is evolving and lacks coordination, so it is not working together to establish standards and popularize, apply and facilitate blockchain technology. Each nation needs to develop a legal framework capable of promoting the adoption and innovation of blockchain products, promoting education in blockchain to harness this lucrative sector's economic potential while protecting consumers and investors from fraudulent practices (Osmani et al., 2020). Briefly, by releasing crypto assets and cryptocurrency, the application promotion environment, the advancement of blockchain technology, and defining responsible entities, the related legal structure mobilizes resources. It is responsible for the monitoring and enforcement of protocols for blockchain (Cermeño, 2016).

12.5 BLOCKCHAIN TECHNOLOGY IN THE BANKING SYSTEM – A TYPICAL CASE STUDY IN A DEVELOPING COUNTRY

12.5.1 Banking Overview in Vietnam

Vietnam's population is more than 96 million individuals, according to the Vietnam Statistical Office (General Statistics Office of Vietnam, 2019), of which 69.3% is the proportion of the population aged 15 to 64. Simultaneously, according to WeareSocial and Hootsuite (2020), approximately 64 million Vietnamese people use the Internet, accounting for 66% of the population in 2019. Also, WeareSocial and Hootsuite (2020) reveals that 52% of people use the Internet to buy online (52% worldwide average) and that the percentage of non-cash purchases is 47% (60% worldwide average). It can be seen that Vietnam is a potential market for the use of information technology goods and services, a fertile piece of cake that gives massive profits to commercial banks (Nguyen et al., 2020).

Following the country's growth, Decree 53/HDBT was issued by the Council of Ministers on 26 March 1988, with a two-level banking structure comprising the State Bank and four commercial banks. After more than 30 years of growth, the number of banks has dramatically developed in terms of quantity and quality: 100% of foreign-owned banks, joint venture banks and credit institutions can participate in the banking industry (Dao et al., 2020; 2021). The rivalry in the banking and finance sectors over globalization and international economic integration is also fierce. Therefore, commercial banks need to diversify goods and services to thrive and expand while providing customers with optimal facilities and applying financial technology. One of the strategic advantages of commercial banks is blockchain technology for transactions between banking products and services.

A milestone of using blockchain in Vietnam occurred when Techcombank pioneered blockchain technology in Vietnam in 2015 for consumer transactions. Techcombank offered to lose an annual remittance fee of 2–3 billion dong and invest USD 30 million in blockchain technology so that customers could make use of modern utilities and free transfer of money inside and outside the bank. The advantages that Techcombank has achieved after 5 years of product and service growth are that it is gaining 5 million customers per year, profit after tax in 2019 is 6.7 times higher than in 2015, and equity in 2019 will rise 3.8 times compared with 2015 (Techcombank, 2020). In addition, the NAPAS (2020), Techcombank was honoured as Outstanding Performance Bank and awarded the "Leading Bank for Interbank Money Transfer Service 24/7 banking".

In 2017, the Vietnamese banking industry began to experience the use of blockchain in money transfer services by three other banks in Vietnam: MBBank, Tpbank and SCB. The introduction of this technology has brought a great achievement: The annual equity is 1.4 times (SCB), 2.3 times (MBBank) and 3.2 times higher (Tpbank) in 2019 compared with 2017. In the same period, profit after tax rose by 8.33% (SCB), 34.74% (MBBank) and 95.82% (Tpbank).

In 2018, using blockchain technology, NAPAS and three commercial banks started to experiment with a money transfer model. For an e-commerce transaction, the average cost is about 1–2%, whereas blockchain-enabled transactions will cost as little as 0.1% or less. The lowest fee reduction promotes the extension of financial services to a wide variety of clients and small businesses towards comprehensive financial targets, according to SupercruptoNews (2019). As suggested in the Vietnam Banks Association (2019), HSBC in 2019 and BIDV as well as Vietinbank in 2020 are formally releasing L/C to introduce blockchain technology. Blockchain technology benefits include high security, the quick processing speed of transactions, reducing paperwork, saving time, restricting errors, allowing parties to complete workflows in real time, and maximizing digital L.C. issuance transactions.

12.5.2 Data and Methodology

The author uses qualitative research methods to evaluate the application of financial technology, especially blockchain, and in-depth interviews are conducted with two objectives. The first objective of the in-depth interview is to define the reality of the application of financial technology in general, especially blockchain, in Vietnamese commercial banks. The second objective is to consider the blockchain trend in banking institutions and financial institutions. Thc writer conducts in-depth interviews with 6 leading experts and 15 employees working in Vietnam's commercial banks. The survey is carried out directly by the survey questionnaire, and through Google Docs, the survey is also online. The author also surveyed young individuals who use banking products and services. The total number of accurate surveys is 62. The survey site is in Binh Duong's province, bordering Ho Chi Minh City, where the southern part of Vietnam is developing dynamically.

12.5.3 Results

An in-depth survey of leaders and employees employed at 14 commercial banks in Vietnam is depicted in Figure 12.4. The proportion of males surveyed in the survey sample is higher than that of females, with the workers being mainly 22–30 years old. Similarly, Figure 12.5 shows the online survey for leaders and employees at 11 commercial banks in Vietnam via Google Docs. The number of men surveyed in the survey sample is higher than the proportion of women, including the director, the head of the department and the workers; the survey age is mainly 22–30 years old. In Table 12.1, the survey findings are summarized: The table synthesizes reviews on blockchain technology trends in Vietnam.

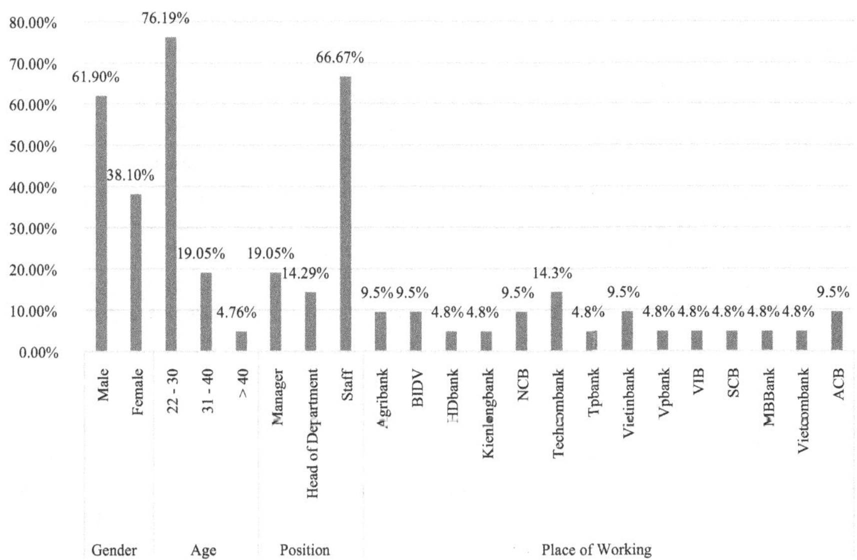

FIGURE 12.4 Gender, age, working position, place of working. (Results from the analysis.)

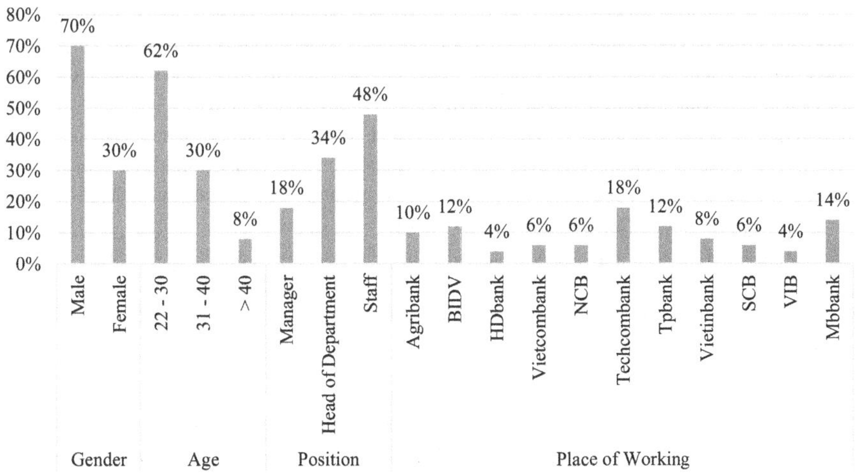

FIGURE 12.5 Gender, age, working position, place of using banking services. (Results from the analysis.)

We conducted a questionnaire in another study on a sample of 62 young people conducting transactions with banking products and services and incorporating financial technology, in particular blockchain, in their transactions, as shown in Table 12.2.

The questionnaire results in Table 12.2 indicate that many individuals use the services at more than one bank in Vietnam. Among them, the most popular are

TABLE 12.1
Blockchain Technology Trends

Comments	Comments on blockchain technology trends in Vietnam
Comment 1	Techcombank – Due to blockchain's use in money transfer services, TCB has achieved excellent business performance. TCB will soon begin to deploy blockchain with many other services and products. Some banks, such as ACB, Sacombank and Vietinbank, will also use blockchain to provide consumers with the best benefits.
Comment 2	Banking is the fastest and most successful business area in which technology is applied. Commercial banks that want to thrive must aim to provide consumers with the best goods and services in the highly competitive market and the explosion of today's Industrial Revolution 4.0. Therefore, an imminent development is the use of blockchain technology in the banking sector.
Comment 3	The bank that goes first would have more benefits in the current technology application race. For example, in Techcombank, many customers have become loyal customers of TCB because of the advantages of blockchain application products and services. It will be difficult for later banks to maintain current clients and grow new ones. Hence, since this is the most advanced technology, they are compelled to select the blockchain framework.
Comment 4	The Covid-19 pandemic has given Vietnamese consumers several opportunities to experience banking services without going to a transaction counter. Vietnamese people have also had plenty of time to compare and select the best banking services at the lowest rate during this period. Although the investment costs for blockchain are very high, it helps to minimize the cost of intermediaries and paper, allowing banks to reduce customs fees and improve the competitiveness of rates.
Comment 5	The goal of the bank is to use blockchain in remittances and cross-border payments. The technology of blockchain aims to reduce intermediaries, shortening transaction times from a few days to a few minutes. Again, not only MBBank but many other banks can also find ways to take advantage of this.
Comment 6	Blockchain is a solution for identifying identity theft – a problem that banks need to better answer. When making purchases, the blockchain network allows consumers to use fingerprints and facial recognition technology. It is understood to be the most appropriate safety step.

Source: Results from the analysis.

the central banks and state-owned banks. In particular, BIDV and MBBank are the banks where the state is the main shareholder, and 80.64% and 79.03%, respectively, of the respondents selected these for most transactions. Sacombank did not receive much attention from young people, unlike private banks and small banks like ACB. The transactions are relatively easy, only depositing and withdrawing cash (100%); evaluating in-depth transactions, the online deposit rate is just 43.5%, and the rate of online loans is exceedingly low, just 4.84%; in addition, banks use financial technology and blockchain in transactions in Vietnam and other emerging economies. The products that users are not very concerned with are also bottlenecks. There is a similar situation for other researchers in India and Africa.

TABLE 12.2
Results of Research

Items		Quantity	Percentage
Gender	Male	11	17.74
	Female	51	82.26
Bank	BIDV	50	80.64
	MBbank	49	79.03
	Techcombank	45	72.58
	Vietcombank	35	56.45
	Tpbank	31	50
	Agribank	20	32.26
	ACB	14	22.58
	Sacombank	7	11.29
Products	Non-cash payment	62	100
	Deposit and withdraw cash	62	100
	Internet banking or smart banking	62	100
	Online savings deposit	27	43.55
	Online lending	3	4.84

Source: Results from the analysis, data collected by the authors.

Based on the survey evaluation, users of banking products and services said they were pleased with the money transfer services of Techcombank and Tpbank because of their quick and cost-free processing speed for transactions. However, with the BIDV service, many customers are not happy because mobile banking sometimes has errors and they cannot access it. In contrast, many people complain that Vietcombank charges very high money transfer fees. It is observed that blockchain technology allows Vietnam's commercial banks to offer the most increased convenience and low cost to clients; growing competition, attracting new clients, increasing profits and increasing the size of equity. However, for many reasons, such as high investment costs, stringent banking regulations and consumer trust, retail banks are still very cautious in adopting blockchain (Higginson et al., 2019).

In contrast to this issue are the attempts to deploy blockchain by governments, investment banks and suppliers. For the big boys like Amazon, Oracle and IBM, blockchain services have become a technology sprint. Hackathons have been launched by wholesale banks and related to fintech. Vietnam officially abolished the household registration book on 1 January 2021, and transitioned to demographic management using technology. Vietnam was ranked eighth in crypto operations worldwide in 2019, according to Hoang (2020). The amount of capital that startups have raised in the Vietnamese blockchain space is approximately $150 million. This is a promising indicator that demonstrates the potential for blockchain growth in Vietnamese commercial banks. It can be argued that the Vietnamese market is a fertile piece of cake in the blockchain space for business operations, and the banking and financial field is no exception. Vietnam is a developing nation with a preference for cash. In 2019, the proportion of Vietnamese individuals with a bank account was about 40%, according to Forbes Vietnam (2020), but 80% of regular purchases still use cash, 98% use cash when paying for products of less than 100,000 dong and 85% account for cash withdrawal transactions at ATMs. The target of the ratio of cash to total means of payment as of 31 December 2019 was 11.33%. A large part of the population preferred online shopping with door-to-door delivery systems and bank transfers to restrict interaction when the Covid-19 epidemic occurred. A study by the State Bank of Vietnam (2020) showed that in the first six months of 2020, there were 200 million Internet payment transactions worth around VND 12.9 million, a rise of 36% in value over the same period in 2019. The number of cell phone transactions was 472 million (1.78 times the number in 2019) with a volume of about 4.9 million dong (equivalent to 1.77 times the 2019 volume). This is considered a significant change for Vietnamese citizens and opens up banks' opportunities to expand the market.

12.6 CONCLUSIONS

The banking industry has consistently delivered innovative technology experiences, such as telebanking, online banking, mobile banking and fintech, to customers in recent years. Business transactions in various areas have been enabled by the application of technology in banking and finance. It is anticipated that blockchain technology will be a breakthrough that will trigger a shift in the shape and scale of the banking and finance sector while at the same time introducing developments in the way business transactions are carried out. With emerging innovations such as blockchain, automation, cloud computing, augmented reality under development, and a significant impact on the banking industry, not just on countries, the Technological Revolution 4.0 is bursting into life in both developed and emerging nations. Experimental data and studies indicate that implementing a blockchain technology framework offers many potential benefits to the financial system and economic growth in developing countries. The advantages of a blockchain, such as transparency, the accuracy of transactions, immutability and high processing speed at low cost, have contributed significantly to the development in the banking system. In addition, blockchain offers the advantage of helping clients trust goods and services and deters transaction fraud. Blockchain's opportunity for banks helps them gain a competitive advantage in the economy's virtualization battle.

In the future, blockchain could also enable banks to build new services. However, there are still some risks to blockchain technology, such as scalability, security risks, interoperability and other issues that need to be addressed. Another possibility, blockchain technology has some challenges, such as operational costs, transaction costs, energy, and storage costs, increasing dramatically. Therefore, before the mass adoption of blockchain technology, banks in developing countries need to carefully analyse these problems to optimize the advantages and minimize the risks that blockchain technology brings.

REFERENCES

Accenture (2017). *Banking on Blockchain: A Value Analysis for Investment Banks*. Dublin, Ireland: Accenture Consulting.

Belinky, M., Rennick, E., & Veitch, A. (2015). The fintech 2.0 paper: Rebooting financial services. Retrieved from http://santanderinnoventures.com/wp-content/uploads/2015/06/The-Fintech-2-0-Paper.pdf

Cermeño, J.S. (2016). *Blockchain in Financial Services: Regulatory Landscape and Future Challenges for Its Commercial Application*. Madrid, Spain: BBVA Research. http://www.smallake.kr/wp- content/uploads/2017/01/WP_16-20.pdf

Chang, V., Baudier, P., Zhang, H., Xu, Q., Zhang, J., Arami, M. (2020). How Blockchain can impact financial services: The overview, challenges and recommendations from expert interviewees. *Technological Forecasting and Social Change, 158*, 120166. https://doi.org/10.1016/j.techfore.2020.120166

Cherukupally, S. (2020). Blockchain technology: Theory and practice. *Handbook of Statistics*. Available at https://doi.org/10.1016/bs.host.2020.10.001

Dao, L.K.O., Nguyen, T.Y., Hussain, S., & Nguyen, V.C. (2020). Factors affecting non-performing loans of commercial banks: The role of bank performance and credit growth. *Banks and Bank Systems, 15*(3), 44–54. http://dx.doi.org/10.21511/bbs.15(3).2020.05

Dao, L.K.O. Loc, H.H., Nguyen, V.C., Hang, L.T.T., Do, T.T. (2021). Factors affecting the choice of banks: Do bank's interest rate, employee image and brand matter? *Journal of Asian Finance, Economics and Business, 8*(1), 457–470. https://doi.org/10.13106/jafeb.2021.vol8.no1.457

Dhar, S., & Bose, I. (2016). Smarter banking: Blockchain technology in the Indian banking system. *Asian Management Insights, 3*(2), 46–53. Available at: https://ink.library.smu.edu.sg/ami/3

Drescher, D. (2017). *Blockchain Basics: A Non-technical Introduction in 25 Steps*. Apress, viewed November 23 2018, ISBN-10: 1484226038.

Forbes Vietnam (2020). 63% of Vietnamese adults having a bank account. Available at https://forbesvietnam.com.vn/tai-chinh/63-nguoi-truong-thanh-co-tai-khoan-ngan-hang-8802.html

Frame, W.S., & White, L.J. (2014). *Technological Change, Financial Innovation, and Diffusion in Banking* (pp. 1–5). New York: Leonard N. Stern School of Business, Department of Economics.

Garg, P., Gupta, B., Chauhan, A.K., Sivarajah, U., Gupta, S., & Modgil, S. (2020). Measuring the perceived benefits of implementing blockchain technology in the banking sector. *Technological Forecasting and Social Change, 159*, 120407. https://doi.org/10.1016/j.techfore.2020.120407

General Statistics Office of Vietnam (2019). Press release on population and housing census results in 2019. Retrieved from https://www.gso.gov.vn/su-kien/2019/12

Hanushek, E.A. (2013). Economic growth in developing countries: The role of human capital. *Economics of Education Review, 37*, 204–212. https://doi.org/10.1016/j.econedurev.2013.04.005

Higginson, M., Hilal, A., & Yugac, E. (2019). Blockchain and retail banking: Making the connection. Available at https://www.mckinsey.com/industries/financial-services/our-insights/blockchain-and-retail-banking-making-the-connection#

Hoang, N. (2020). Blockchain calls for regulatory sandboxes and education. Available at https://www.rmit.edu.vn/news/all-news/2020/feb/blockchain-calls-for-regulatory-sandboxes-and-education

HSBC (2020). HSBC performing the first letter credit transaction on blockchain platform in Vietnam. Available at https://www.about.hsbc.com.vn/news-and-media

Kshetri, N., & Voas, J. (2018). Blockchain in developing countries in I.T. Professional. *IEEE*, 20(2), 11–14. doi: 10.1109/MITP.2018.021921645.

MacDougall, W. (2014). *Germany Trade and Invest: Industrie 4.0: Smart Manufacturing for the Future.* Berlin: Gesellschaft für Außenwirtschaft und Standortmarketing mbH. Available at http://www.gtai.de/GTAI/Navigation/EN/Invest/Service/publications,did=917080.html

Mainelli, M., & Smith, M. (2015). Sharing ledgers for sharing economies: An exploration of mutual distributed ledgers (aka blockchain technology). *Journal of Financial Perspectives*, *3*(3), 2–44.

Manikandan, A. (2019). *ICICI, Kotak, Axis among 11 to Launch Blockchain-linked Funding for SMEs.* Availble at Economicstimes.

McKenzie, B. (2018). Blockchain and cryptocurrency in Africa A comparative summary of the reception and regulation of Blockchain and Cryptocurrency in Africa. Retrieved from https://www.bakermckenzie.com/-/media/files/insight/publications/2019/02/report_blockchainandcryptocurrencyreg_feb2019.pdf

Mehdiabadi, A., Tabatabeinasab, M., Spulbar, C., Karbassi Yazdi, A., & Birau, R. (2020). Are we ready for the challenge of banks 4.0? Designing a roadmap for banking systems in industry 4.0. *International Journal of Financial Studies*, *8*(2), 32.

Mendling, J., Weber, I., Aalst, W.V.D., Brocke, J.V., Cabanillas, C., Daniel, F., ... & Gal, A. (2018). Blockchains for business process management-challenges and opportunities. *ACM Transactions on Management Information Systems (TMIS)*, *9*(1), 1–16.

Morgan Stanley (2018). Counterpoint global insights: Blockchain. Available at https://www.morganstanley.com/im/publication/insights/articles/ii_theedgeblockchain_us.pdf

Mosconi, F. (2015). *The New European Industrial Policy: Global Competitiveness and the Manufacturing Renaissance*. London: Routledge.

Nakamoto, S. (2009). Bitcoin: A peer-to-peer electronic cash system. Available at https://bitcoin.org/bitcoin.pdf

NAPAS (2020). Techcombank honoring the outstanding performance bank. Retrieved from napas.com.vn

Nguyen, T.T., Nguyen, V.C., Tran, T.N. (2020). Oil price shocks against stock return of oil and gas-related firms in the economic depression: a new evidence from a copula approach. *Cogent Economics & Finance*, *8*(1), 1799908. DOI:10.1080/23322039.2020.1799908

Osmani, M., El-Haddadeh, R., Hindi, N., Janssen, M., & Weerakkody, V. (2020). Blockchain for next generation services in banking and finance: cost, benefit, risk and opportunity analysis. *Journal of Enterprise Information Management*, *34*(3), 884–899.

People's Bank of China (2017). *Annual Report.*

Peters, G.W., & Panayi, E. (2016). Understanding modern banking ledgers through blockchain technologies: Future of transaction processing and smart contracts on the Internet of money. In *Banking Beyond Banks and Money* (pp. 239–278). Cham: Springer International Publishing.

Santander (2019). Santander launches the first end-to-end blockchain bond. Retrieved from https://www.santander.com/en/press-room/press-releases/santander-launches-the-first-end-to-end-blockchain-bond

State Bank of Vietnam (2020). *Annual Reports.* Available at sbv.gov.vn

SupercruptoNews (2019). Blockchain and Vietnam's banking activities. Retrieved from https://www.supercryptonews.com/blockchain-and-vietnams-banking-activities

Techcombank (2020). *Annual Reports.* Retrieved from https://techcombank.com.vn/nha-dau-tu/bao-cao-thuong-nien

The Bank of Thailand (2020). *Annual Report.*

Tyson, L., & Lund, S. (2016). Digital globalization and the developing world. Available at https://www.project-syndicate.org/commentary/digital-globalization-opportunities-developing-countries-by-laura-tyson-and-susan-lund-2016-03?barrier=accesspaylog

UNDP (2020). Human development report. Retrieved from http://report.hdr.undp.org

University of Pennsylvania (2018). How the Blockchain brings social benefits to emerging economies. Available at https://knowledge.wharton.upenn.edu/article/blockchain-brings-social-benefits-emerging-economies/#:~:text=Developing%20countries%20such%20as%20India,peer%20mechanism%20for%20verifying%20information.

Vietnam Banks Association (2019). Applying Blockchain in the bank in Vietnam. Retrieved from http://www.vnba.org.vn/index.php?option=com

Vovchenko, N.G., Andreeva, A.V., Orobinskiy, A.S., & Filippov, Y.M. (2017). Competitive advantages of financial transactions on the basis of the blockchain technology in digital economy. *European Research Studies, 20*(3B), 193–212.

WeareSocial and Hootsuite (2020). Vietnam digital 2020. Retrieved from https://eliteprschool.edu.vn/vietnam-digital-2020-bao-cao-cua-hootsuite-wearesocial/

World Bank (2017). Blockchain in financial services in emerging markets part II: Selected regional developments. Retrieved from https://www.ifc.org/wps/wcm/connect/f12930a4-a78b-43c0-9fe1-35018cdd9fb5/EMCompass+Note+44.pdf?MOD=AJPERES&CVID=m4Qcm5N

World Bank (2019). Blockchain : Opportunities for private enterprises in emerging markets: Chapter 4. In *Blockchain in Financial Services in Emerging Markets—Current Trends* (2nd ed.). Washington, DC: International Finance Corporation. https://openknowledge.worldbank.org/handle/10986/31251 License: CC BY-NC-ND 3.0 IGO.

World Bank (2020). Trade (%GDP). Available at https://data.worldbank.org/indicator/NE.TRD.GNFS.ZS

World Economic Forum report (2016). *Why Trust in the Digital Economy is Under Threat.* Avaiable at http://reports.weforum.org/digital-transformation/building-trust-in-the-digital-economy/

13 Decentralizing Finance
Cryptocurrencies, ICOs, STOs and Tokenization of Assets

Hazik Mohamed

CONTENTS

13.1 INTRODUCTION: BACKGROUND AND DRIVERS

There are several key drivers that are making decentralized finance work – tokenization of real-world assets, maturity of stable coins, and the improved acceptance of regulations and standardization. We are witnessing two fast-growing trends merge and complement each other. The first one is tokenization, where all illiquid assets in the world, from private equity to real estate and luxury goods, become liquid, and all liquid assets can be traded more efficiently. The second is the rise of a new tokenized economy where inevitably, new rules will develop, categorized as tokenomics. The new economy is enabled by solving the volatility in cryptocurrencies with the introduction of stablecoins (Mohamed, 2020b). By unlocking the economic potential of blockchain, these two complementary and correlated trends

DOI: 10.1201/9781003138082-13

will complete the decentralization of finance and the way financial services of smart cities of the future will be implemented.

In this chapter, the sectoral disruptions within the financial management and services industry that we discuss include cryptocurrencies, capital markets, and asset and portfolio management, and we provide an example on tokenizing agriculture and livestock. The identified DeFi[1] applications for financial management and services are in the tokenization of assets of value and the subsequent processes that involve client onboarding, financial prediction, management of model portfolios, payment systems, trade clearing and settlement (Mohamed, 2020a). We begin with the fundamentals of money and the evolution of cryptocurrency issuance from initial coin offerings (ICOs) to security token offerings (STOs) and then to stablecoins like central bank digital currencies (CBDCs). We then examine the concept of tokenizing assets and deliberate on a conceptualization of a tokenized capital markets trading platform. To further clarify tokenization and its many benefits, we provide an example on the tokenization concept for agriculture and livestock in raising capital for small farms. As in all digital transformation concepts, we include discussions on concerns for privacy and security, which emphasizes self-governance, self-regulation and cybersecurity measures.

The financial management and services industry plays a critical role in our economy, such that it supports all economic activities through capital formation, preservation and distribution. However, it suffers from a trust deficit as a result of persistent financial crises due to unethical and non-performance of fiduciary duties by entities that have been entrusted with such duties. Technology was seen as a viable way to automate or make operationally inherent such checks and balances that would improve the governance of financial services. The fallibility of human judgement, or the lack of it, makes digital transformation more crucial when a code of ethics and corporate governance fails to live up to its noble expectations. Fraud and abuse detection mechanisms can help enforce accountability and security in financial services.

Digital transformation is also important to the financial services industry for competitiveness and stability in the markets. The sustainability of any organization relies on its ability to innovate and embrace change and new ways of doing business to scale and substantially enhance efficiency and elevate performance within the organization. Without the reinvention and transformation of existing processes, services cannot reach the evolving levels of financial efficiency and rising utility of the people who use such products and services.

13.2 CONCEPTUAL FRAMEWORK

The conceptual framework investigates and proposes various use cases for the application of artificial intelligence (AI) and blockchain to support the range of financial services and the essential underlying services associated with financial institutions and their customers. Our discussion intends to demonstrate the right use of each technology to bring out its benefits when applied appropriately and to understand the relationships of technologies and how they can be used to serve the shift in consumer behaviour. The paradigm shift brought on by digital transformation is altering the

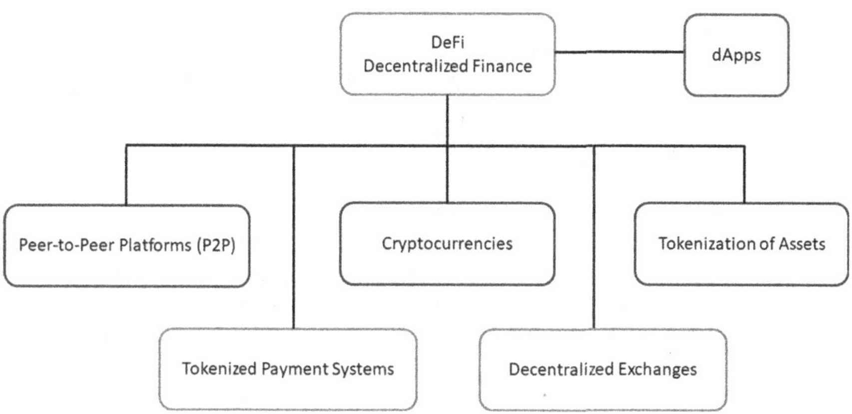

FIGURE 13.1 Broad categorization of decentralized finance. (Author's own.)

customers' behaviour towards financial services and how it is being managed and deployed. Today, consumers want financial services at the tips of their fingers from anywhere, at the tap of their mobile screens, wherever they are – in offices, at home or in a park. Mobility, instant gratification and shorter attention spans are now priority considerations as financial management and services go mobile and agile.

As such, our discussion covers a broad spectrum of areas (see Figure 13.1) from the onboarding of customers to financial platforms, advisory via chat-bots, AI-driven financial predictions from historical data with respect to forward-looking market data, decentralized payment systems, digitized trade settlements, portfolio construction and asset management, and the overall framework for decentralized finance to work sustainably.

13.3 MONEY, FIAT CURRENCY AND CRYPTOCURRENCIES

The three basic purposes of money are as a unit of account, a medium of exchange and a store of value. Cash is a physical asset with the following features: (i) It is anonymous, (ii) it is universal, (iii) it is exchangeable from peer to peer (deprived of issuer details) and (iv) it does not yield any interest on its own (Mohamed, 2020). Banks are the traditional money makers and preserve their individuality at maintaining reserves at the central banks (CBs). Peer-to-peer interchangeability permits its exchange among counter-parties without mediators. Universality means that anyone can take ownership of it, utilize and save it. Historically, currencies were backed by value, which can be done via gold, silver, commodities or the government. In fact, fiat money used in the current economic system is considered a currency, and this includes the emergent use of plastic and digital currencies (digital sovereign fiat). Its core function is to serve as a customary accepted form of payment by the people in exchange for possessions and facilities.

Cryptocurrencies are a form of virtual currencies built on blockchain technology, which may be native or Bitcoin derived, and the first cryptocurrency appeared in

2009. "Cryptographic methods are at the middle of their execution" (He et al., 2016), and factually, the idea and perception of storing significant information by utilizing cryptography techniques is considered to be much older, as the word "crypto" is from an ancient Greek term *kryptos*, meaning "hidden". In some records, it is stated that "ancient Egyptians also made use of cryptography as it is evinced by practicing cipher by Julius Caesar in 100BC to 40BC" (Fry, 2018).

Cryptocurrencies bear some similarities to regular currencies. Unlike regular currencies, cryptocurrencies are purely digital assets, supported by blockchain-enabled encryption techniques and using cryptography for securing transactions, regulating the formation of extra elements and authenticating the transmission of resources. The critical difference between the two is that cryptocurrency can be created independently of CBs and can be used independently of typical regulated financial intermediaries (such as banks). Unlike sovereign authorized currencies, cryptocurrencies are not legal tender (i.e., guaranteed by government). Forms of money and the categorization of cryptocurrencies are shown in Figure 13.2.

Crypto tokens are a form of cryptocurrency that may appear as typically as equity, security or utility tokens, whose purchasing power and right of exchange is limited to a specific asset, product or service for which the token is issued. Additional kinds of tokens consist of "asset-backed tokens" – tokens that represent a physical asset such as real estate, "vote tokens" – tokens that confer on their holder the right to be involved in a project development and "hybrid tokens" – tokens that are a hybridization of more than one form of tokens or their representation.

Cryptocurrencies are championed by their advocates due to their ability to conduct strictly peer-to-peer exchange without intermediaries like banks involved. Their critics are concerned with their being used in illicit transactions, as the system is essentially decentralized and formally unregulated. Another problem for cryptocurrencies is their volatility. As such, the most recent phenomenon is the creation of stablecoins.[2] Likewise, the CBs of major economies have begun to rethink their own

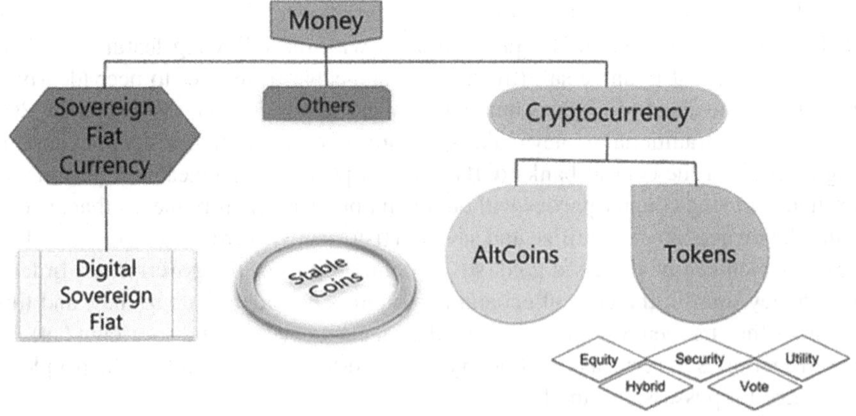

FIGURE 13.2 Forms of money and the categorization of cryptocurrencies. (Author's own.)

sovereign fiat currencies to leverage the benefits of tokenizing currencies in terms of a more efficient interbank settlement system (Mohamed, 2020). Such a legal-tender central bank-issued digital currency is called a central bank digital currency (CBDC), whose value is pegged to its sovereign fiat currency value.

13.4 FROM ICOS TO STOS

An ICO is an innovative form of raising capital or investments by issuing tokens or alternative cryptocurrencies, that does not necessarily involve any equity being acquired by the token buyers or investors. ICOs act to raise funds, whereby a company can raise money via tokenization of its business venture. Investors in ICOs are typically speculators who expect the tokens to skyrocket in value and are therefore only relying on the team behind the project to improve the value of the tokens.

Unfortunately, ICOs are often likened to stocks or shares in the venture. In effect, many of the tokens issued are more like a utility to be exchanged for a product or utilized as a service. Unlike stocks, ICOs do not grant the token-holder any right to equity or profit-sharing of the company's revenue. ICOs were a duplication of the IPO (initial public offering) without having built the venture to a level of maturity that would represent some form of substantial value from revenue performance. Instead, they were meant to fast-track the venture's ability to raise capital based just on the strength of its idea and the team behind that idea. Needless to say, this raised many concerns, especially in an unregulated environment.

Then, in 2018, after several mishaps and complaints, the SEC "delivered a recommendation (continuously updating) citing that every Coin Offerings are security tokens. Numerous controllers trailed suit soon after, this led to the expansion of an innovative form of Coin Offerings labelled the STOs or Security Token Offerings" (SEC, 2019). Still utilizing the tokens issued as a representation of an investment, the STO security token now signifies a "contract into a basic investment asset, like stocks, funds and real estate investment trusts (REIT)" (see Figure 13.3). Like a typical security, it is a "fungible, accessible financial gadget which embraces a kind of monetary value, or an investment product which is sponsored by a real-world asset like a corporation" (SEC, 2019).

The additional requirements included that the venture must prove that it is viable through proof of data, proof-of-concept (POC)/minimal viable product (MVP) or prototype, and other elements like traction to show evidence of viability.

Unlike the ICO market, in which two-thirds of projects are futile or are found to be scams, the STO market had advanced survival and success rates. The necessary supervisory and lawfully enforceable necessities levied on STOs deterred entities with duplicitous intentions from using the DeFi movement for their dishonest purposes. Looking ahead, such steps are necessary if this channel of securitization is to be an option for mainstream adoption.

Also, it is critical to distinguish between private securities and public securities. If a token is considered and handled solely as a private security, in many jurisdictions, there will be the possibility of using certain regulatory exemptions. In these cases, the comprehensive IPO requirements will only come into force if a private token

FIGURE 13.3 Fundamental differences between ICOs and STOs.

(STO) violates the restrictions related to the exemptions in question and begins to act as a public security (see Table 13.1).

But an advantage of an STO over an IPO is that the security token can behave like a programmable share and offers a set of functionalities and attributes that a traditional public IPO share does not have. For instance, an STO can be created into an investment product that is better than traditional stocks (and bonds) by way of rewarding investors both from the forefront, through a risk-sharing model to share profit (and losses), and from the back, through dividend payments. STO companies that can offer profits from the front and back of the business simultaneously have the potential to become the new darlings of profit-driven investors.

13.5 TOKENIZATION OF ASSETS OF VALUE

Tokenization is likely to become one of the most important and influential trends in the crypto space in the coming years. As mentioned earlier, tokens are a representation for "something" on the blockchain, which does not necessarily have to be a currency – like Bitcoin – but could also be a wide range of other types of tangible or intangible assets.

By tokenizing private securities, we can potentially map out and transform illiquid costly assets to highly liquid assets with higher cost-effectiveness. Given that the asset categories within the private securities market are in the trillions of dollars (where the illiquidity discount can be as high as 20–30%), the tokenization of private securities has the potential to unlock billions of dollars in value.

The basic idea of tokenization is the use of smart contracts on a blockchain to create a virtual representation of a certain asset in the form of a token. Depending on the type of asset to be tokenized, different tokens and token standards have been developed for the tokenization process, and different challenges and

TABLE 13.1
Important Distinctions between IPOs, ICOs and STOs

IPO	ICO	STO
IPO gives you ownership of the company based on the number of shares acquired.	ICO give rights of project, not the company equity.	STO tokens represent a share of an underlying asset.
Financial data according to exchange of IPO issued.	As outlined within the white paper and investor agreement.	A security offering under the qualification of an investment contract.
Subject to taxes, with investors liable to capital gains tax.	ICO company may not be taxed; investor subject to capital gains tax.	Subject to taxes, with investors liable to capital gains tax.
An IPO is a onetime sale with multiple intermediaries.	ICOs can have multiple rounds with no intermediaries, the white paper as the blueprint.	STOs have limited intermediaries (lawyers, advisors, no bankers).
Stock exchanges and companies listed by IPO are heavily regulated.	ICO exchanges are not regulated.	STO are somewhat regulated.

opportunities come with the tokenization of the asset in question. Tokenizing tangible real-world commodities differs from tokenization of intangible assets like a software licence. Tokenizing fungible assets like identical types of shares differs from the tokenization of non-fungible assets like a unique work of fine art (see Figure 13.4). Regardless of the type of asset to be tokenized, the basic purposes and benefits are the same: By tokenizing assets and thus equipping them with a virtual representation in the form of a token on a blockchain, it is possible to cut away costly and inefficient middlemen for decentralized trade and exchange, which is faster and easier. Developing blockchain and smart contract-based tokenization platforms will create solutions for tokenization of all kinds of real-world assets, from intellectual rights to commodities to collectables to real estate, with the purpose of increasing liquidity, cutting costs, enabling fractional ownership of assets and opening up the estimated US$280 trillion market of real-world assets for investment. This makes it possible for anyone, anywhere in the world, to invest and create a future global investment market far more democratized than the market of today.

13.6 THE TOKENIZED CAPITAL MARKETS AND DEBT MARKETS

A fully developed capital market is made up of a primary market, where securities are created, and a secondary market, where those securities are traded. The most common capital markets are the stock exchanges, where financial products such as equities (e.g. ownership shares such as stocks) are traded, as well as the bond market, where interest-bearing debt instruments are bought and sold.

FIGURE 13.4 Tokenizing different types of assets. (Author's own.)

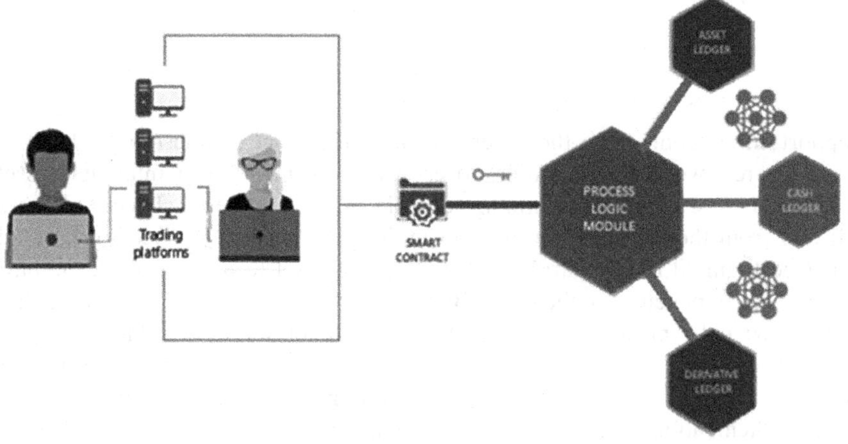

FIGURE 13.5 Conceptualization of a tokenized capital markets trading platform. (Author's own.)

With tokenized systems on the blockchain, the capital market trading platform can be streamlined (see Figure 13.5), provided that an automatic trade lifespan where all parties in the transaction would have access to the exact same data about a trade. This leads to extensive infrastructural cost reserves, effectively managing data, quicker processing cycles, negligible settlement and the potential elimination of brokers and mediators overall (see Figure 13.6). In the DeFi model of capital markets, the primary market where securities are created uses STOs, and the secondary market where these securities are traded will use crypto-exchanges. The ownership of shares and securities is tokenized and recorded by smart contracts for safekeeping or trading.

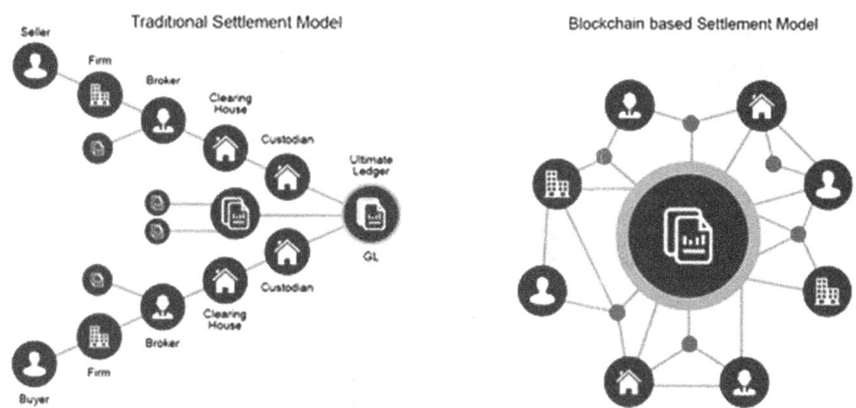

FIGURE 13.6 Comparative settlements in capital markets. (Author's own.)

As for the bond market, digitalization will reduce the cost of origination and issuance, allowing the quantum for digital bonds to be much lower and hence, enabling issuances in the range of US$1–10 million. This opens up the debt capital market space for smaller corporations, which have lower revenues and assets and are unable to issue traditional bonds for capital. They typically have to resort to high-interest corporate loans, which does not really help them to securitize any of their assets. Through tokenization, they will be able to unlock liquidity from their illiquid but valuable possessions. Such digital issuances can be further strengthened by risk-sharing asset-backed concepts like the *sukuk* (Mohamed, 2021).

Furthermore, a digital bond can be a structure for social impact, such that the payment structure keeps "accountability a requirement for the delivery of goods and services, and that effectively spreads the risk of the underlying project amongst the investors and shareholders, instead of transferring the entire risk to the issuer without having achieved the objectives" (Mohamed, 2019).

13.7 EXAMPLE: UNLOCKING VALUE IN AGRICULTURE AND LIVESTOCK

As per the United Nations' Food and Agriculture Organization (FAO, 2015), "about 2/3s of the emerging world's 3 billion rural people live in about 475 million small farms works on land smaller than 2 hectares" (Nagayets, 2005). A large majority are poor and in constant need of food and clean water. Access to basic markets, finance and essential services is limited. Since their options are hampered, they end up farming the land and harvesting food from nature. They eventually become skilful farmers, rearing cattle and many forms of livestock and growing crops, which collectively provide for an extensive quantity of the population worldwide. However, as the years go by, capital and credit constrain their ability to adequately use natural resources for maximum production. Their lack of credit, and subsequent scoring information, inhibits their access to capital in many ways and severely impacts the availability of rural agricultural financing.

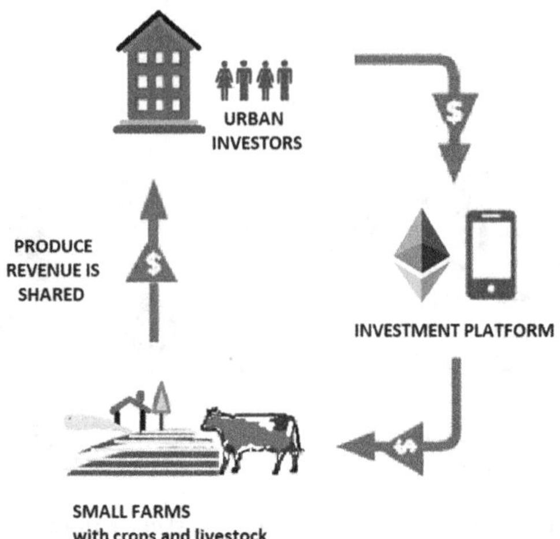

FIGURE 13.7 Tokenizing agriculture and livestock for capital investment. (Author's own.)

An innovative investment platform can be built to provide access to capital for these farming households from urban white-collar workers who are looking for a better investment platform. Livestock can be viewed as valuable assets that can be tokenized to unlock capital to fund production, land, labour and other critical resources required to ensure effective farming output (see Figure 13.7). However, livestock presents an unconventional form of collateral for financing. As such, special provisions need to be in place before investors may want to risk their investments. Among those provisions are the health tracking and ability to insure these mortal and roaming assets (livestock and farm animals) to ensure the successful production cycle for both the farmers and urban investors. With the auditability and traceability of the tokenized (blockchained) system involving mobile phone apps, embedded wireless sensors on the livestock, and an online platform that monitors the progress of each batch cycle, value can be distributed directly to investors who wish to support the production of meat, poultry and other forms of agriculture for small farms. Such traceability to monitor and audit makes insurance coverage possible for these unique assets, which will greatly help in mitigating risks in an unfortunate event that affects the livestock, to circumvent the loss of income for the farmers.

13.8 BENEFITS OF TOKENIZATION ON SECURITIES MANAGEMENT AND TRACEABILITY

One of the most obvious benefits of tokenization is that it will enable liquidity for illiquid assets. The short-lived run of mortgage-backed securities (MBS) was an example of illiquid asset securitization, but its deceptive ambiguity was unravelled when we understood how they were packaged. The consequences of the deceptively securitized MBS resulted from components of the financial product that were not investment-grade (mixed with investment-grade assets), which began to deteriorate, consequently impacting the overall portfolio. The main issue was that although securitization of MBS brought liquidity, its underlying exposure in the repackaged products could not be traced. Traceability of tokens fixes this problem, as the connection to all the underlying assets with a repackaged product is clear and tamper-proof. Thus, the exposure of all assets can be determined even if the product is erroneously or fraudulently rated as a AAA investment-grade product. If rating agencies fail to carry out their duties, the innovative repackaging of financial products to create more liquidity remains possible via tokens, as the underlying exposure is transparent and can be diversified accordingly without being misdirected by asymmetric or false rating information.

Tokenization will also unleash new business opportunities in the area of custody, the safekeeping of real assets, wealth and investment advisory. In particular, the impact on the asset management industry is projected to be quite significant, especially in terms of portfolio construction. Assets under the investible universe would traverse beyond traditional equities or bonds. Asset managers of this new asset class will not only have a new opportunity set for portfolio construction but will need to hone their skills in fundamental research, price discovery and investment-related views to determine what price to pay versus the value the investor gets in return. Ultimately, the core task of active management and its importance will be in demand in this new space. Part of the new capabilities that asset managers need to recruit or acquire are direct real estate expertise, valuation of works of art expertise, and patent and technology expertise, among others, in order to accurately determine the value of the underlying asset, assess its market price and eventually inform an investment decision.

13.9 SELF-GOVERNANCE, SELF-REGULATION AND CYBERSECURITY

Blockchain-based frameworks can be addressed as the management of identity, privacy and security across regions and platforms by decentralizing and distributing organization. Such dispersed platforms provide integrated systems like an identity and the ability to make and "receive payments, pass in composite agreements and conduct without an intermediate" (Mohamed & Ali, 2019).

The only way to comfort acquiescence loads is to construct and organize financial and investment management resolutions by means of blockchain. The ability of blockchain to authenticate and verify information through a consensus mechanism provides a trusted way to identify persons or parties, which makes the entire transaction reliable and trustworthy. Hence, a financial and investment management system on the basis of cryptography can be developed via anti-money laundering (AML), counter-terrorism financing (CTF) and know-your-client (KYC) requirements as per the rules set in each region or country. The whole explanation is focused on the distributed ledger, where an enterprise is a node in the blockchain network, and the platforms developed by asset management companies are driven by an AI engine that checks on AML, CTF and KYC compliance and builds a database of records. Computerized audits, programmed reporting and process restructuring are other aids offered by such AI and blockchain platforms to address regulatory compliance, where technology is assisting financial service providers with regulatory requirements and compliance management.

At the same time, the "wide usage of internet has shaped concerns of loss of personal data, information leaks or loss of money by cyberattacks" (Mohamed & Ali, 2021). The pattern now is that cyberattacks are planned and executed in a sophisticated way in order to achieve their criminal goals. In order to preserve value and safety for usage at scale, all companies must take extensive steps to get protection from cyberattacks, particularly in the storage and safekeeping of tokens and other critical information. Cloud security is probably one of the most important avenues to prevent cybersecurity attacks. Typically, fintech companies use cloud-based services to scale their products and services with lower running costs.

> Maintaining integrity of markets is very significant for increasing tokenization possibilities, hence detecting criminal cyber actions helps in overcoming the question of data breaches concerning service companies amongst others separately from emerging the legal national framework to allow the local law-enforcement activities to collaborate with law-enforcement agencies overseas.
>
> **(Mohamed & Ali, 2021)**

For any financial scheme or monetary system to be viable or sustainable, it must be protected by a framework of laws and enforcement mechanisms to protect the rights of individuals and ensure the stability of the market system. Enforcement of laws involves the detection of legal violations and the subsequent appropriate penalty to discourage everyone from violating the guidelines, beyond relying on trust alone.

Unfortunately, defending against hackers is similar to defending against an invisible enemy or a deadly virus. All investors must ensure that not only the ICT departments or selected people have responsibility; rather, it's the responsibility of all staff, corporations, institutions and government. It only takes a single weak link to start the breach of token attacks or unwanted spread of private information.

13.10 CONCLUSION

By agreeing the classification of assets and ownership and performing activities on more than half of a particular asset, distributed ledger technology/blockchain allows greater revenue. By reducing investment barriers, a wide range of people can purchase/finance in goods. In conventional markets, these technologies can help traders find partners more easily for making transactions. They also support complete financing by the initial investment market to a wide range of stakeholders. Without the requirement for an intermediary function, investors are now able to access investment prospects where their participation has been restricted due to infrastructure reasons or very high investment constraints. Now, access to the financial markets and a variety of new types of assets have been permitted irrespective of the investor's location, and there are very low cost requirements. Assets that are subdivided incorporate the concept of shared ownership, whereby more people can buy property together and use it.

In decentralizing finance, money as a means of exchange can be viewed as exchanging fragments of information of value. There is no doubt that cryptographic breakthroughs that will revolutionize financial transactions are being developed as we write and read about them. The technology and innovation advancements will pave the way for a revolution in every transaction, beyond financial transactions only. Among the important developments that need to take place, it needs to be understood how the price discovery of varied cryptocurrencies translates to returns, and its attendant volatilities, as well as the impact of volatilities from additional assets (including other cryptocurrencies, tokenized assets, stocks, merchandises and bonds, amongst others). Such in-depth understanding will be crucial to develop hedging and diversification strategies for financial management in investment portfolios or treasury operations.

13.10.1 Limitations of Research

As much of the research focused on the viability of a decentralized peer-to-peer blockchain-based platform, the bigger picture of macroeconomic benefits was not scrutinized. It would be beneficial to also understand how the alternative finance of the DeFi movement in terms of volume per capita has an impact on a country's GDP performance. It would be interesting to have empirical data showing that the growth of DeFi initiatives may make a significant contribution to financial inclusion and efficiencies that positively impacts the GDP of a nation.

13.10.2 Recommendations for Future Research

While this research highlights the benefits of utilizing peer-to-peer blockchain-based platforms for various use cases, it also does not ignore the reality of the barriers to blockchain platforms. Future research could discover the emergent initiatives in this area and conduct an in-depth examination into Uniswap, MakerDAO, Compound, Aave, Yield protocol, dYdX, Synthetix and so on. One should always be cautious when embracing the possibilities and opportunities with their associated risks. Future research should indeed explore whether any major risk factors may develop out of the DeFi space over the next few years, and how we can deal with them.

NOTES

1. DeFi is a terminology short for "decentralized finance" for a variety of financial applications in frontier technologies geared toward disrupting financial intermediaries.
2. Stablecoins are cryptocurrencies designed to minimize their price volatility by pegging to some "stable" asset or basket of assets, such as fiat money and commodities (like precious metals or industrial metals).

REFERENCES

Food and Agriculture Organization of the UN (FAO). (2015). The Economic Lives of Smallholder Farmers, from http://www.fao.org/family-farming/detail/en/c/385065/

Fry, J. (2018). Rise of Crypto Exchanges and why it is important. Retrieved August 1, 2018, from https://www.linkedin.com/pulse/rise-crypto-exchanges-why-important-jonny-fry/

He, D., Habermeier, K.F., Leckow, R.B., Haksar, V., Almeid, Y., Kashima, M., & Kyriakos-Saad, N. (2016, January). Virtual Currencies and Beyond: Initial Considerations. Retrieved from http://www.imf.org/external/pubs/cat/longres.aspx?sk=43618

Mohamed, H. (2019). Blockchain-based Impact Sukuk. In Hidayat, S.E., Aryani, Y.F., Fianto, B.A., Rusmita, S.A., & Nisful Laila, H. (Eds.), *Blending Islamic Finance and Impact Investing for the SDGs* (Chapter 6, pp. 106–123). Republic of Indonesia: Fiscal Policy Agency, Ministry of Finance.

Mohamed, H. (2020). Implementing a Central Bank Issued Digital Currency: Assessing Economic Implications. *International Journal of Islamic Economics and Finance*, *3*(1), 51–74.

Mohamed, H. (2021). *Beyond Fintech: Technology Applications for the Digital Economy.* Singapore: World Scientific.

Mohamed, H., & Ali, H. (2019). *Blockchain, Fintech and Islamic Finance: Building the Future of the New Islamic Digital Economy.* Boston/Berlin: DeIG Press.

Mohamed, H., & Ali, H. (2021). Finding Solutions to Cybersecurity Challenges in the Digital Economy. In Boitan, I.A., & Marchewka-Bartkowiak, K. (Eds.), *Fostering Innovation and Competitiveness with FinTech, RegTech, and SupTech* (Chapter 5, pp. 80–96). Pennsylvania, PA: IGI Global.

Nagayets, O. (2005). Small Farms: Current Status and Key Trends. Paper prepared for the Future of Small Farms Research Workshop, Wye College, U.K. June 26–29, 2005.

Prasad, A., & Sethi, J. (2019). *Message Authentication; Emerging Security Algorithms and Techniques* (pp. 249–272). https://doi.org/10.1201/978135102170

Securities Commission (SEC). (2019). Framework for "Investment Contract" Analysis of Digital Assets. from https://www.sec.gov/corpfin/framework-investment-contract-analysis-digital-assets

Suhel, S.F., Shukla, V.K., Vyas, S., & Mishra, V.P. (2020, June). Conversation to Automation in Banking through Chatbot Using Artificial Machine Intelligence Language. In 2020 8th International Conference on Reliability, Infocom Technologies and Optimization (Trends and Future Directions) (ICRITO) (pp. 611–618). IEEE.

14 Role of Blockchain Technology in Digital Forensics

Keshav Kaushik, Susheela Dahiya and Rewa Sharma

CONTENTS

14.1 INTRODUCTION TO BLOCKCHAIN TECHNOLOGY

Blockchain technology is gaining popularity in the present time due to its properties, such as peer-to-peer transaction, persistency, anonymity, immutability and decentralized behaviour. The concept of blockchain emerged from the bitcoin cryptocurrency proposed by Satoshi Nakamoto (Nakamoto, n.d.). It is a distributed and decentralized digital ledger having a consensus of various economic transaction schemes. The term *bitcoin* was coined in the year 2008; it was the earliest cryptocurrency, which later gave rise to more than 2000 cryptocurrencies. The usage of bitcoin is not adopted worldwide because of various issues like illegal mining performance (Dev, 2014), anti-money laundering (Moser et al., 2013) and the time consumed in the transaction validation and mining process (Battista et al., 2015; Beikverdi & Song, 2015; Wijaya, 2017). In this digital world, unreliable transmission channels are often used to transfer information from one point to another. Blockchain technology can be applied where any digital data requires to be stored, executed, verified or updated among several users with transparency, privacy, trust and security, without control by any central authority. Blockchain technology offers reliable peer-to-peer communication, thus ensuring the security and privacy of users. Once a transaction is recorded, it can be read publicly but can't be modified. Blockchain can be specified

DOI: 10.1201/9781003138082-14

as a probabilistic state machine and cannot be used in domains where exact decisions are required (Saito & Yamada, 2016). This technology can be used for solving various issues in several domains, such as Internet of Things (IoT), government identification, healthcare, insurance, shipping, border control, energy, real estate, advertising, waste management, supply chain management, industry and many more such application areas.

The properties that make blockchain technology different from traditional databases are as follows:

14.1.1 Distributed: Blockchain is not a centralized scheme and has a peer-to-peer network technology. All the incoming data blocks can be verified, recorded, stored and updated by all other nodes in the blockchain. If the data is found valid, it is added to the chain; if not, it is discarded.

14.1.2 Autonomous: It is the consensus property of the blockchain system that makes it autonomous in nature. Every node on the system can perform the desired operation on a data block with the consensus of the majority of nodes present in the network.

14.1.3 Transparent: Any kind of update or addition in any block by any node of the blockchain is visible to all the nodes of that blockchain, thus improving the transparency of the system.

14.1.4 Immutable: Blockchain blocks are preserved for the future, and any illegal modification in any block can be done only after gaining the computational powers of 51% of all the nodes present in the network.

14.1.5 Anonymous: Most blockchain systems are open to the public and support anonymous transactions, only requiring the user to know the blockchain wallet address.

14.2 APPLICATION AREAS OF BLOCKCHAIN TECHNOLOGY

The idea of blockchain technology emerged from bitcoin cryptocurrency, but later, the concept was used for different applications. Some of the application areas of blockchain are discussed here:

14.2.1 Healthcare: Better services for end users can be provided by the healthcare sector adopting the latest technologies, such as blockchain. Health data related to patients is crucial for individuals, families and society. Privacy is one of the major concerns that need attention in the healthcare sector. Blockchain can be used to ensure the privacy of the digital database, which can be shared among various users.

14.2.2 Internet of Things: IoT can be defined as a collection of various heterogeneous devices exchanging information with each other. The application domain of IoT includes smart agriculture, smart healthcare, wearable technology, smart retail, smart cities, smart homes, smart grid, smart healthcare, etc. Most of the IoT applications have centralized system architecture. Some of the drawbacks of having a centralized system are single point failure, security issues and trust management. Shala et al. (2019) elaborated the trust evaluation consensus protocol for digital ledger in Machine-to-Machine (M2M) systems. The integration of IoT and blockchain technology makes the system tamper proof and more robust. Filho et al. (2019) and Carnevale et al. (2019) described energy-efficient distributed computing in a Body Area Network. In Aly et al. (2019), details of layers of IoT applications and various security issues are explained.

14.2.3 Legal Perspective: Blockchain offers a decentralized platform for the distribution of digital content. The licensor gives permission to upload each digital document using the smart contract. Users can also access the digital documents available in the blockchain database using the smart contract. Giancaspro (2017) and Nath (2016) explained that the present legal frameworks can use blockchain as an effective way to maintain a digital ledger without having any involvement of a third party, thereby ensuring privacy and security of individuals.

14.2.4 Government: There are several potential advantages of implementing blockchain in the government sector, such as sharing of information across various organizations in the country, accessibility, transparency, and quality and quantity of service enhancement. Sullivan and Burger (2017) and Nordrum (2017) elaborated how blockchain technology can be implemented in the government sector due to its properties of being secure, peer to peer, decentralized and accessible to all users in the network.

14.2.5 Advertising: Blockchain is quite suitable for digital advertising due to its inherent characteristics of being immutable, transparent, and a distributed digital ledger. Inefficiency, lack of transparency and advertisement frauds are some major concerns faced by the advertising sector. Blockchain technology can potentially help to boost efficiency, improve transparency and optimize costs, and can significantly reduce the number of frauds. MetaX (Goldin Mike et al., 2017), a blockchain company, utilizes blockchain technology to deal with the issues of transparency and fraud affecting digital advertising.

The applications of blockchain are not limited to these areas but span various domains such as digital forensics, smart transportation, reputation systems, cloud computing, the commercial world, e-business, insurance, etc.

14.3 CHALLENGES AND PROTOCOLS OF BLOCKCHAIN TECHNOLOGY

14.3.1 Challenges Faced by Blockchain Technology

One of the major challenges faced by blockchain technology is scaling up the system, as it is not as fast as other electronic transaction schemes. For example,

MasterCard and Visa cards can perform thousands of transactions per second, whereas a bitcoin blockchain can process around ten transactions per second (Kosba et al., 2016). So, to design an efficient and distributed blockchain system is quite a challenging issue.

One more concern that needs to be addressed is to design an efficient algorithm to take consensus of all nodes, ignoring the illegitimate nodes, in a permission-based system.

The mining process involved in blockchain systems requires a lot of computation power, so it is the need of the hour to design an effective mining process that can be executed with the requirement for less computation power.

The blockchain system offers a lot of advantages and possesses such attractive characteristics that everyone wants to become part of this network. This also adds to the responsibilities of the developers. In such a scenario, when the blockchain grows in size, efficient mechanisms need to be employed to store and verify data by each node.

Other challenges that also affect blockchain systems are replacement of current databases, handling interoperability issues, effectively implementing distributed task scheduling, etc.

14.3.2 Blockchain Architecture and Protocols

Blockchain technology allows a global database to be accessed from anywhere by anyone having an internet connection. A blockchain is not owned by any central party like banks or government. It is nearly impossible to cheat the system when the entire network is looking after the blockchain, thus protecting users from fake documents and fraudulent transactions.

Figure 14.1 displays the basic architecture of a blockchain system, where users are connected with each other in a distributive manner. Various operations that a user can perform are initiating and validating a transaction or mining process. Information is stored permanently among the nodes in the network. A local copy of the blockchain system is stored at each node of the network, which is periodically updated to ensure consistency among all the nodes of the network. A blockchain is a distributed information-sharing platform, which allows multiple nodes that do not trust each other to take part in the decision-making process. As the system is decentralized, it doesn't suffer from the single point failure issue.

Figure 14.2 displays various types of blockchain. Presently, it can be classified as public, private and consortium blockchain. In public blockchain, the ledger is made public to all the users on the internet. Transactions can be added and verified to the blockchain by anyone on the network (Xu et al., 2017). On the other hand, private blockchain allows the addition and verification of transaction blocks only by specific users of the organization, but anyone on the internet can view the ledger (Dinh et al., 2017). In consortium blockchain, only specific organizations can add or verify a transaction block.

The basic element of the structure is a block. A block comprises a group of valid transactions. In a blockchain system, any user can start the transaction and broadcast it to all the users present in that network. New transactions can be validated using

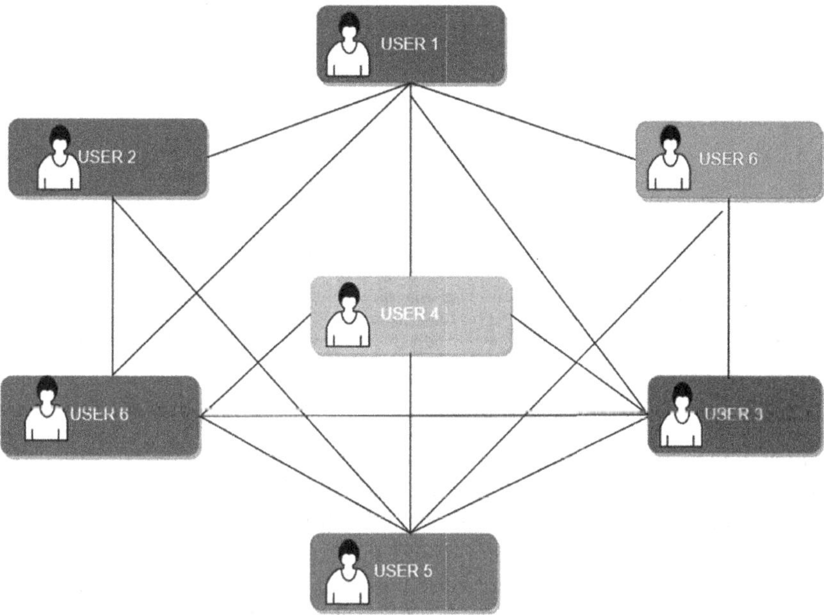

FIGURE 14.1 Basic architecture of blockchain system.

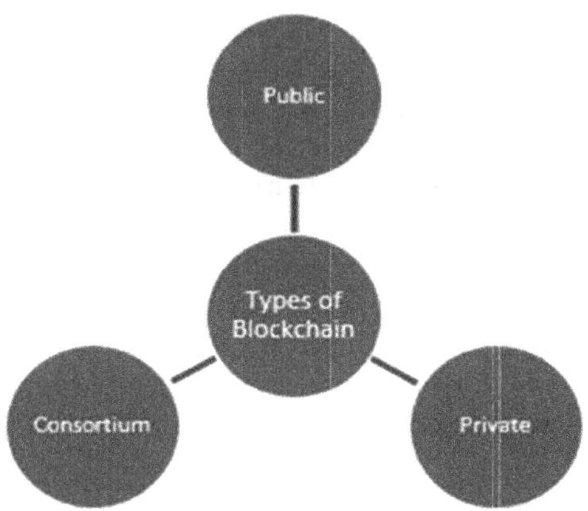

FIGURE 14.2 Types of blockchain.

old transactions, and after validation, the transaction is added to the existing blockchain. Transactions are made part of a particular block according to the time frame in which they occurred. In bitcoin,

> "A block may contain more than 500 transactions on average, the average size of a block is around 1 MB (an upper bound proposed by Satoshi Nakamoto in 2010)".
>
> **http://satoshinakamoto.me/whitepaper/**

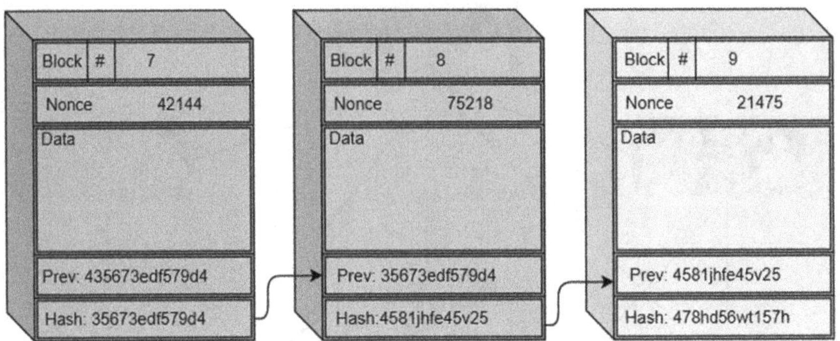

FIGURE 14.3 Attributes of a block in blockchain.

A larger number of transactions can be processed simultaneously with a larger block size. Block attributes of a blockchain are described in Figure 14.3.

Data is stored in these blocks in the chain. Hashing algorithms are used for varying-length transactions to produce a fixed-length output and broadcast it to all the users present in the network. In order to make the chain tamper proof, the previous block's hash is used to create the new block's hash. The Merkle tree function is used by bitcoin blockchain to generate a final hash value. This value is stored as the hash of the current block, and each succeeding block stores the Merkle root of the previous block.

In order to avoid centralization, blockchain does not depend on atomic clocks such as NTP (Network Time Protocol). So, each block contains a timestamp, which signifies the block generation time. When a new block is received by a node, it checks that that timestamp specified in the block doesn't exceed UTC time by more than 100 ms. Blocks also store additional information such as signature, nonce value and other user-defined data. Blockchain runs on a network of computers, which stores all the data and updates in the blockchain. These computers are known as miners. These miners perform the Proof of Work mechanism to achieve consensus.

Consensus function is a method of keeping the database consistent. A new transaction is validated only after the agreement of each blockchain node. The most common consensus mechanism used by both bitcoin and Ethereum is Proof of Work (PoW).

Proof of Work – A PoW mechanism is used to generate a piece of data that is quite difficult to create but can be easily confirmed by others, satisfying specific requirements. Difficulty is computed considering the computation power, electricity and time consumption. PoW generation is a random process having low probability and requires a lot of trial-and-error attempts before generating a valid PoW. The Hash Cash PoW system is used by bitcoin. The process of calculating this PoW data is called "mining". The "Nonce" value is a random value stored in the block headers. PoW requires the Nonce value of the blocks to be exchanged in order to produce a value that makes the hash value in the block header lower than a preset "difficulty target". Miners must generate a PoW that covers entire data blocks. It is quite unpredictable which node in the network will generate the new block due to the low chances of successful generation.

Proof of Stake – As mentioned earlier, PoW generation consumes a lot of time, computing power and electricity. On the contrary, Proof of Stake (PoS) doesn't require high computing power. In the PoS mechanism, the amount of bitcoin a miner possesses is compared with the resource. For example, only 1% of the PoS blocks can be mined by a miner possessing 1% of the bitcoin. This mechanism is useful in preventing malicious attacks on the network, as it's quite costly to execute an attack, and the attacker himself suffers from his own attack.

14.4 MANAGING CHAIN OF CUSTODY FOR DIGITAL EVIDENCE: ETHEREUM

In cyberfraud detection, digital data plays an important role because it connects individuals to illegal activity. Therefore, it is critically important to ensure that digital information is complete, genuine and auditable as it passes through various layers of hierarchy in the custody chain during forensic cybercrime. The capability of blockchain technologies to allow an interconnected view of transactions back to their origin gives the forensic community tremendous hope. In today's modern world, the value of digital documentation for people connected to cybercriminals is also rising exponentially. Digital evidence poses its own special chain of custody (CoC) challenges. Custody chain can be described as a mechanism for preserving and recording the history of digital documentation. Proof from digital forensics moves from the first responder to higher authority in charge of managing cybersecurity analysis across various layers of hierarchy. Forensic Chain (Lone & Mir, 2019) is a blockchain-based custody chain maintenance and traceability solution. Blockchain is a data system that generates a digital recording and storage history (events/records) that is exchanged across the public computing network by all participating parties. Blockchain uses encryption to secure the recording and recording operations (events/records) in the network, thereby providing an audit trail.

With regard to the CoC, blockchain capabilities may generate documents concerning access to tamper-proof evidence, especially in combination with cryptographic hacking and encryption. The proof to be stored is first safely crypted and has blockchain power built on. The cryptographic data will only be available to the desired party on the blockchain but would simultaneously record the accessing party's time, date and likely user ID and apply it to the blockchain. The blockchain is read in a manner that resembles how the bitcoin blockchain can be decoded by a special feature. This blockchain feature enables courts and relevant workers to study the historical custody chain without requiring access to data themselves.

The following are the benefits of using blockchain in the digital forensics domain:

- Proof may be obtained, stored and checked.
- Where the mechanism was initially entered may track the route of any incident or operation.

- It also leads to enhancing the transaction efficiency and cost savings of transactions of certain kinds by increasing transparency, thus removing the need for a trustworthy third party to verify certain statements and to pass facts and proof of confidence based on consensus, resulting in greater trust between communicating parties.
- The record itself, allowing proven proofs that can both be confirmed.
- Fraud avoidance due to improved audit trail disclosure.
- It helps organizations to integrate proof into the verification of the case or event.

14.5 MANAGING CHAIN OF CUSTODY FOR DIGITAL EVIDENCE: HYPERLEDGER

In every digital forensic analysis, CoC has a significant role to play, because during its transit across multiple layers of hierarchy, it tracks any minute data on digital information from the first respondent to the higher authority responsible for handling the investigation of cybercrime. CoC documents facts like how, where, when and who came into contact with the data etc. evidence was collected, analyzed and stored for development. However, forensic CoC can be corrupted if it is not stored and preserved adequately over the lifespan of digital proof, making proof of cybercrime evidence inadmissible to the court of law. For the facts to be admissible in the case, it is also necessary to maintain the credibility of the digital judicial CoC.

CoC refers to the recording process and to the maintenance of the chronological history of digital proof management. To defend CoC from unwanted modification or damage, exceptional caution is required. The ultimate objective of CoC is to prove that the accusations are genuinely true for the suspected crimes rather than evidence being wrongly planted. Poor CoC results in digital testimony being inadmissible in court. Blockchain's use case creation and depletion procedures are relatively straightforward, and the production period is shortened dramatically from months to weeks. One of Hyperledger Composer's many benefits is that it is entirely free, with a model of open governance that encourages anyone to contribute to the project. Hyperledger Composer stands on and supports current Hyperledger Fabric blockchain technology and running time to ensure that the transactions are verified in compliance with the policy specified by the designated business network members using the pluggable blockchain consensus protocol.

Hyperledger Caliper (GitHub – Hyperledger/Caliper: A Blockchain Benchmark Framework to Measure Performance of Multiple Blockchain Solutions Https://Wiki.Hyperledger.Org/Display/Caliper, n.d.) is a blockchain performance appraisal standards system and is one of several Linux Foundation host Hyperledger ventures. Caliper enables users with a predefined series of software to evaluate the performance of various blockchain implementations and to produce reports with a set of performance checks. The Caliper reports include a series of performance metrics such as second transfers, transaction latency, resource usage, etc. Caliper's existing supports are Cloth, Sawtooth, Iroha and Composer-based blockchain solutions. Hyperledger works Caliper in several steps. The planning process begins with the implementation of intelligent contracts to plan a test environment. The second process (Lone & Mir, 2019) consists of checking the execution of tasks delegated to customers to execute predefined test cases based on a transaction count or on a length basis. The final step is a performance review period in which all the test reports are obtained for documentation. The management of digital information raises unprecedented difficulties because it is latent, unpredictable and unstable and can rapidly, efficiently and sometimes in a time/machine-dependent way cross jurisdictional boundaries. Ensuring that the systems and methods used to obtain and pass evidence in a digital society are authentic and legalized is also a real challenge. Digital data's hash code, the location of the criminal scene and officers' identities are no longer enough to allow facts to be taken to court. It is therefore important to recognize all the parties involved with the forensic investigation and provide the precise location of each item of digital evidence and the persons who have access to evidence and have a full list for all transactions, such as the digital signature of each product. Moreover, the forensic scientist depends on electronic forensic equipment, and the reliability of test findings is thus highly dependent on whether those techniques are accurate and how they are implemented.

The following is the flowchart shown in Figure 14.4 for the systematic process followed in maintaining the CoC with the help of Hyperledger:

Phase 1 – Generation of Evidence: In this phase, the evidence is generated a unique ID, which is generated by taking the hash value of that evidence, which keeps track of the integrity. Other characteristics, such as the author and host, are often set to the first participant address/ID. The participant's address/identification will be transferred, showing that it is the developer and the first owner of the digital evidence. The production period of proof is also mentioned in this phase.

Phase 2 – Evidence Transmission: The evidence transmission process takes the proof ID and address as input and transfers the possession in return to the given address. This phase first tests whether facts exist, and if so, it sets the proof owner to the new individual, who invokes the task.

Phase 3 – Evidence Elimination: The evidence elimination feature takes the evidence ID as input and deletes the necessary proof. It first tests whether there are any facts, and whether or not it excludes the evidence from the database, the participant who invokes it is the source of the evidence. No participant may withdraw specific evidence, but a contract can be provided stating that such information is no longer applicable to a specific event.

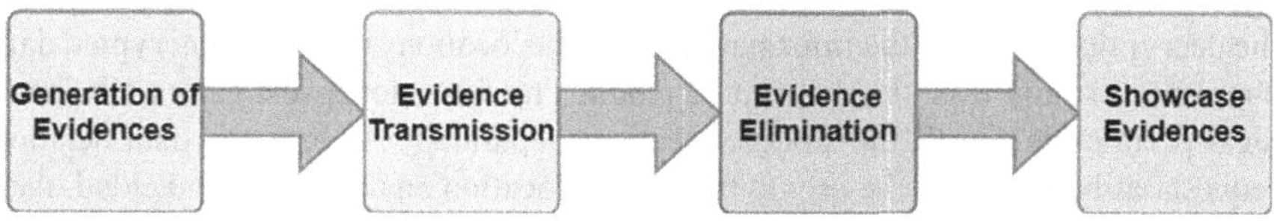

FIGURE 14.4 Process for maintaining CoC in Hyperledger.

Participants' documentation and replicas of the original digital evidence, on the other hand, are primarily regarded. The current proofs are kept in a safe place initially.

Phase 4 – Showcase Evidence: The showcase evidence function uses evidence input ID and returns the blockchain proof information. The only measure is to ensure that the proof exists now.

14.6 APPLYING BLOCKCHAIN FOR DISTRIBUTED CLOUD STORAGE IN DIGITAL FORENSICS

With the advancement of technology, the volume of web data generated on a daily basis is increasing tremendously. To handle this vast volume of data, most enterprises, application developers and consumers are choosing a cloud storage environment. A centralized cloud storage environment like Google cloud, ownCloud and Dropbox provides the facility to access data anytime, anywhere. In centralized cloud storage, the cost of data communication over the cloud is very high. Also, preserving the security and privacy of private and sensitive data on the cloud has become a critical issue because along with the use of the cloud storage environment, the security threat to the data has also increased. The recovery of data as well as collecting digital forensics is also very difficult for the forensics investigators because it involves access to every bit of private and confidential data stored on the central cloud (Ricci et al., 2019).

Along with centralized cloud storage, a new concept of distributed cloud storage is emerging that uses blockchain technology, which is by default encrypted. In distributed cloud storage, the unused storage of one user can be offered on rent to another user for storing their data. In this way, the user can take the unused storage space of other users on rent preferably in his/her own geographic proximity. All users of distributed cloud storage are connected through a peer-to-peer network. This network is more secure, faster and less expensive than centralized cloud storage. The blockchain technology in distributed cloud storage ensures the confidentiality and integrity of the data and provides verifiable temper-proof data without any middlemen. In blockchain-based distributed cloud storage, everything that happens to the data, like its different storage locations, details of owner and details of the person who accesses that data, has been monitored. All these details of the data can be fetched by anyone who has access to the blockchain. Despite this, the collection of digital forensics on distributed cloud storage becomes more complicated. If the data is not available or recoverable (in the case of deletion) from local storage, then the forensics investigators need to recover the data and metadata from the distributed cloud storage, which requires login credentials for getting access to the account, along with the decryption keys and the information of all the locations where the encrypted data chunks are stored over the distributed cloud. The forensics investigators also need legal permission for all the storage locations separately. Also, if the data deletion request has been sent by the user to the storage location on the distributed cloud, then after the deletion of that data chunk, it will be impossible to determine the location where those data chunks existed, given that there will be no metadata existing that can indicate the deletion. All these challenges make the acquisition of data from blockchain-based distributed cloud storage more complex. A lot of research has been required for developing the methods/applications required for recovering data from blockchain-based distributed cloud storage for digital forensics.

14.7 CONCLUSION

This chapter has provided a brief introduction to blockchain technology along with its unique properties that make it different from traditional databases. After that, its different application areas, like healthcare, IoT, government, advertising and legal perspectives, have been discussed. Then, the challenges faced by blockchain technology, its architecture and the protocols have been discussed. After that, the process of managing CoC for digital evidence (preserving and recording digital documentation historical history) in Ethereum and Hyperledger has been discussed in detail. Finally, the chapter describes the difference between centralized cloud storage and blockchain-based distributed cloud storage along with the advantages and disadvantages of applying blockchain for distributed cloud storage. The inclusion of blockchain in distributed cloud storage preserves the integrity and confidentiality of data, but the process of collecting digital forensics becomes more complex.

REFERENCES

Aly, M., Khomh, F., Haoues, M., Quintero, A., & Yacout, S. (2019). Enforcing security in Internet of Things frameworks: A systematic literature review. *Internet of Things*, 6, 100050. https://doi.org/10.1016/J.IOT.2019.100050

Battista, G. Di, Donato, V. Di, Patrignani, M., Pizzonia, M., Roselli, V., & Tamassia, R. (2015). Bitconeview: Visualization of flows in the bitcoin transaction graph. In 2015 IEEE Symposium on Visualization for Cyber Security, VizSec 2015. https://doi.org/10.1109/VIZSEC.2015.7312773

Beikverdi, A., & Song, J. (2015). Trend of centralization in Bitcoin's distributed network. In 2015 IEEE/ACIS 16th International Conference on Software Engineering, Artificial Intelligence, Networking and Parallel/Distributed Computing, SNPD 2015 - Proceedings. https://doi.org/10.1109/SNPD.2015.7176229

Carnevale, L., Celesti, A., Galletta, A., Dustdar, S., & Villari, M. (2019). Osmotic computing as a distributed multi-agent system: The body area network scenario. *Internet of Things*, *5*, 130–139. https://doi.org/10.1016/J.IOT.2019.01.001

Dev, J.A. (2014). Bitcoin mining acceleration and performance quantification. *Canadian Conference on Electrical and Computer Engineering*. https://doi.org/10.1109/CCECE.2014.6900989

Dinh, T.T.A., Wang, J., Chen, G., Liu, R., Ooi, B.C., & Tan, K.-L. (2017). BLOCKBENCH: A framework for analyzing private blockchains. In Proceedings of the ACM SIGMOD International Conference on Management of Data, Part F127746 (pp. 1085–1100). https://arxiv.org/abs/1703.04057v1

Filho, G.P.R., Villas, L.A., Gonçalves, V.P., Pessin, G., Loureiro, A.A.F., & Ueyama, J. (2019). Energy-efficient smart home systems: Infrastructure and decision-making process. *Internet of Things*, *5*, 153–167. https://doi.org/10.1016/J.IOT.2018.12.004

Giancaspro, M. (2017). Is a 'smart contract' really a smart idea? Insights from a legal perspective. *Computer Law & Security Review*, *33*(6), 825–835. https://doi.org/10.1016/J.CLSR.2017.05.007

GitHub - hyperledger/caliper: A blockchain benchmark framework to measure performance of multiple blockchain solutions https://wiki.hyperledger.org/display/caliper. (n.d.). Retrieved July 14, 2021, from https://github.com/hyperledger/caliper

Goldin, M., Soleimani, A., & Young, J. (2017). *The AdChain Registry.*

Kosba, A., Miller, A., Shi, E., Wen, Z., & Papamanthou, C. (2016). Hawk: The blockchain model of cryptography and privacy-preserving smart contracts. In Proceedings of the 2016 IEEE Symposium on Security and Privacy, SP 2016 (pp. 839–858). https://doi.org/10.1109/SP.2016.55

Lone, A.H., & Mir, R.N. (2019). Forensic-chain: Blockchain based digital forensics chain of custody with PoC in hyperledger composer. *Digital Investigation, 28*, 44–55. https://doi.org/10.1016/J.DIIN.2019.01.002

Moser, M., Bohme, R., & Breuker, D. (2013). An inquiry into money laundering tools in the Bitcoin ecosystem. *ECrime Researchers Summit, ECrime.* https://doi.org/10.1109/ECRS.2013.6805780

Nakamoto, S. (n.d.). *Bitcoin: A Peer-to-Peer Electronic Cash System.* Retrieved July 14, 2021, from www.bitcoin.org

Nath, I. (2016). Data exchange platform to fight insurance fraud on blockchain. In IEEE International Conference on Data Mining Workshops, ICDMW (pp. 821–825). https://doi.org/10.1109/ICDMW.2016.0121

Nordrum, A. (2017). Govern by blockchain Dubai wants one platform to rule them all, while Illinois will try anything. *IEEE Spectrum, 54*(10), 54–55. https://doi.org/10.1109/MSPEC.2017.8048841

Ricci, J., Baggili, I., & Breitinger, F. (2019). Blockchain-based distributed cloud storage digital forensics: Where's the beef? *IEEE Security and Privacy, 17*(1), 34–42. https://doi.org/10.1109/MSEC.2018.2875877

Saito, K., & Yamada, H. (2016). What's so different about blockchain?-Blockchain is a probabilistic state machine. In Proceedings of the 2016 IEEE 36th International Conference on Distributed Computing Systems Workshops, ICDCSW 2016 (pp. 168–175). https://doi.org/10.1109/ICDCSW.2016.28

Shala, B., Trick, U., Lehmann, A., Ghita, B., & Shiaeles, S. (2019). Novel trust consensus protocol and blockchain-based trust evaluation system for M2M application services. *Internet of Things, 7*, 100058. https://doi.org/10.1016/J.IOT.2019.100058

Sullivan, C., & Burger, E. (2017). E-residency and blockchain. *Computer Law & Security Review, 33*(4), 470–481. https://doi.org/10.1016/J.CLSR.2017.03.016

Wijaya, D.A. (2017). Extending asset management system functionality in bitcoin platform. In Proceeding of the 2016 International Conference on Computer, Control, Informatics and Its Applications: Recent Progress in Computer, Control, and Informatics for Data Science, IC3INA 2016 (pp. 97–101). https://doi.org/10.1109/IC3INA.2016.7863031

Xu, L., Shah, N., Chen, L., Diallo, N., Gao, Z., Lu, Y., & Shi, W. (2017). Enabling the sharing economy: Privacy respecting contract based on public blockchain. In Proceedings of the ACM Workshop on Blockchain, Cryptocurrencies and Contracts. https://doi.org/10.1145/3055518

15 A Study and Experimentation on Bitcoin Price Forecasting with Time Series Analysis and Recurrent Neural Network (RNN)

Martin Aruldoss, V. Adarsh and Miranda Lakshmi Travis

CONTENTS

15.1 INTRODUCTION

Bitcoin is considered a widely recognized cryptocurrency in the digital world and is one of the well-recognized investment assets (Chen et al., 2020c). Cryptocurrency provides a platform to transfer currency from one end to another end safely without any identification (anonymously). It has created an impact on the global economic system, and this is not managed by any government or any centralized agencies. It is mobilized digitally through cryptography and peer-to-peer protocols (Jalali and Heidari, 2020). Many investors started to think about bitcoin as one of the very safest investments compared with gold, platinum, safety bonds and so on. However, during the pandemic situation resulting from the first and second waves of COVID-19, the assumptions about bitcoin being the safest asset failed because of its drastic fluctuations. Also, many queries have been posted on Google about the safety of this decentralized investment (Chen et al., 2020a).

DOI: 10.1201/9781003138082-15

The price of bitcoin has been influenced by many factors. Bitcoin price and investors attention are very closely related, it has been influenced by information-driven and noise components (Ibikunle et al., 2020). Another interesting fact about bitcoin has been identified: The close relationship between bitcoin users, bitcoin miners and investors, and bitcoin price. These factors strongly influence the bitcoin price (Chen et al., 2020b). In most cases, bitcoin is considered a speculative asset or investment because of its rapid growth and sudden fluctuations. A quantitative framework has been developed to predict bitcoin price, bitcoin price behaviour, identification of strongly influential parameters, the sentiment of bitcoin users, investor interest, and attention towards investment and selling. This model is a continuous-time model, and it did not consider bitcoin as currency but rather, as a speculative asset because of rapid changes and fluctuations in price (Cretarola et al., 2020).

The speculative nature of bitcoin has created many prediction models using different kinds of techniques to predict the price of bitcoin. Statistical techniques (Khedr et al., 2021), machine learning techniques (Jalali and Heidari, 2020), recurrent unit approach (Dutta et al., 2020), grey system theory, ensembles of neural networks (Sin & Wang, 2017), deep learning techniques (Ji et al., 2019), Auto-Regressive Integrated Moving Average (ARIMA) (Poongodi et al., 2020) and other prediction techniques have been applied to predict bitcoin price. This research proposes a hybrid model in which recurrent loops in the activation function of the recurrent neural network (RNN) model are combined with ARIMA.

In this chapter, Section 15.2 describes a literature review on stochastic models and RNNs, Section 15.3 describes the research problem, Section 15.4 describes experimentation and data sets used for bitcoin price prediction, Section 15.5 describes the results with implications and Section 15.6 concludes the chapter.

15.2 LITERATURE REVIEW

The price of bitcoin changes every day, and the change can be considered as a daily price change as well as a high-frequency price change. The bitcoin price change is dominated by many factors such as property and network, trading and market, attention and gold spot price. These factors are considered high-dimensional features. Also, the basic bitcoin trading feature exchange of cryptocurrency is considered to predict the bitcoin price for a short interval of time (Chan et al., 2020c).

The price of bitcoin is affected by many factors (Chan et al., 2020c), and it is measured using two phases. In the first phase, the different trends are associated with bitcoin, and in the second phase, the available information about bitcoin is considered to measure the price using machine learning techniques (Velankar et al., 2018). Similarly, bitcoin price is predicted using a decision tree in which parameters such as bitcoin's open price, high price, low price and closing price are considered (Rathan et al., 2019). The bitcoin price is also compared with time series analysis and machine learning techniques (Felizardo et al., 2019).

Also, deep learning techniques such as long short-term memory (LSTM) and gated recurrent unit (GRU) are applied to predict the bitcoin price. These techniques were found to be very good at predicting the price of bitcoin very accurately

(Awoke et al., 2021). Sequence learning algorithms are widely applied for time series forecasting, and LSTM is one of the sequence algorithms. It has not been applied specifically to cryptocurrencies like Bitcoin, Ethereum, Binance Coin, Dogecoin, XRP, Tether, Cardano and so on. Hence, a research model has been proposed to predict the bitcoin price with LSTM and with a variation in LSTM called AR (2). This model performed well compared with other models in predicting bitcoin prices very accurately (Livieris et al., 2021).

A survey has been conducted to analyse the impact of different bitcoin price prediction techniques such as statistical techniques, machine learning techniques and deep learning techniques. The research results indicate that compared with statistical techniques (which require more assumptions), machine learning techniques perform well in predicting bitcoin price (Khedr et al., 2021).

Bitcoin price is also influenced by many factors such as demand raised for bitcoin, the amount of bitcoin in circulation, and the exchange value of bitcoin. These parameters are applied to predict the price of bitcoin using time series analysis methods like GRU, ARIMA, and LSTM (Gupta & Nain, 2021). The volatile nature of bitcoin takes time series analysis as a very important model for price prediction. Hence, an ensemble of time series models combined with a neural network has been developed for bitcoin price prediction. Two kinds of shifts have been found in bitcoin prices: Deterministic and moderate. According to a time interval (for example day, hour, minute, etc.,), the price change is decided and the shift is stated as a moderate or inevitable (Shin et al., 2021).

From the literature, it has been found that bitcoin is very volatile, and its price changes very quickly. Moreover, it is one of the decentralized currencies. Hence, predominant models are required to predict the price of bitcoin very quickly and more precisely. An alert system should be developed with machine learning when the threshold level of bitcoin price is reached (Shankhdhar et al., 2021).

A special kind of research has been conducted to investigate the different prediction techniques available to predict the bitcoin price. This research has emphasized that predicting bitcoin price with higher accuracy is very important, as it is essential for investors for investment and good profit-making. Also, it is essential for policymakers to make policy decisions and for researchers to better understand the financial market situation and investments. This research has identified that many statistical models, time series analyses, machine learning techniques and deep learning techniques are applied to predict the price of bitcoin. When these results are compared with real bitcoin price analytics, a huge difference is identified. Hence, the research suggests that more sophisticated and specialized models and algorithms need to be developed to conduct bitcoin price analytics (Pintelas et al., 2020).

15.3 ARIMA WITH RECURRENT NEURAL NETWORK FOR FORECASTING

Time series analysis is one of the popular methodologies applied for the prediction of future performance from past performance. From the literature, it is evident that even with many stochastic models in this game for bitcoin price prediction, it was an absolute challenge to forecast with high accuracy. And moreover the bitcoin price prediction is a linear time-bound property.. Hence, a better model has been proposed with times series analysis and artificial neural network methodology for bitcoin prediction.

The ARIMA model is one of the well-recognized time series models, and it has been applied in many domains to identify future price predictions. For example, it has been applied in domains such as stock price prediction, electricity prices, supply chain, traffic interval prediction (Lv et al., 2021), chiller system performance (Ho & Yu, 2021), COVID-19 – probable evolution of the pandemic (Benvenuto et al., 2020; Alaraj et al., 2021), wind speed forecasting (Liu et al., 2021; Kavasseri & Seetharaman, 2009), forecasting primary fuel demand (Ediger and Akar, 2007) and so on.

The hybrid model of ANN (Artificial Neural Network) and ARIMA has outperformed all the other forecasting time series models (Guo et al., 2021), (Patel et al., 2020) for bitcoin price prediction. This model has been designed from the advantages of linear and non-linear modelling. Moreover, the different kinds of ARIMA models produce results with minimal errors; however, the error did not make much difference with actual data (Ismail et al., 2021). The linear model ARIMA, combined with backpropagation neural network (BPNN) has shown better prediction accuracy compared with other combinations of linear and non-linear models (Yang et al., 2021). The performance of short-term models are compared with neural networks for forecasting in the supply chain process.. In this analysis, the neural networks performed well compared with short-term models (Bousqaoui et al., 2021). Moreover, research has been conducted on stock price prediction using neural networks with ARIMA. In this research, the performance of ARIMA is improved when it is combined with the ANN models (Kinasz, 2021).

The method used in this research is a hybrid model, which is introduced by sandwiching the ARIMA model with RNN, which, it is hoped, will improve the accuracy over the current prediction, since this method is very similar to the RNN and LSTM method. The results obtained by the ARIMA model with RNN and RNN with LSTM are evaluated to find the performance and efficiency. In this research, the input stream is first forwarded into the ARIMA model, which generates a relatively good and accurate result. This result is again passed into the gates of the RNN model for its neural convolutions under 300 epochs, which further subdivides and classifies to produce a better-valued prediction. The proposed model is expected to create a better result with low resource consumption as well.

15.4 EXPERIMENTATION

For the experiment, a data set was collected from Kaggle for the years 2017 and 2018 (Kaggle, 2021). The details of the data set are as follows:

- coincheckJPY_1-min_data_2014-10-31_to_2018-01-08.csv
- bitflyerJPY_1-min_data_2017-07-04_to_2018-01-08.csv
- coinbaseUSD_1-min_data_2014-12-01_to_2018-01-08.csv
- bitstampUSD_1-min_data_2012-01-01_to_2017-01-08.csv

Attributes count and Attribute's list:

Eight numerical attributes are selected, as follows:

- Timestamp.
- Open.
- High.
- Low.
- Close.
- Volume_(BTC).
- Volume_(Currency).
- Weighted price.

Class count and class list: In this experiment, continuous numerical values are predicted. The Instance count: A total of 3,161,057 instances approximately selected for training and testing. The different steps involved in this experimentation are data preprocessing, important feature selection and loading training, and testing data loaded with neural network and backpropagation network is applied to minimize the error rate. The results are discussed in the next section.

15.5 RESULTS AND DISCUSSION

The results obtained using two models, RNN with ARIMA and RNN with LSTM, are depicted as a graph in Figure 15.1 and Figure 15.2 at different points of time (a particular date is considered).

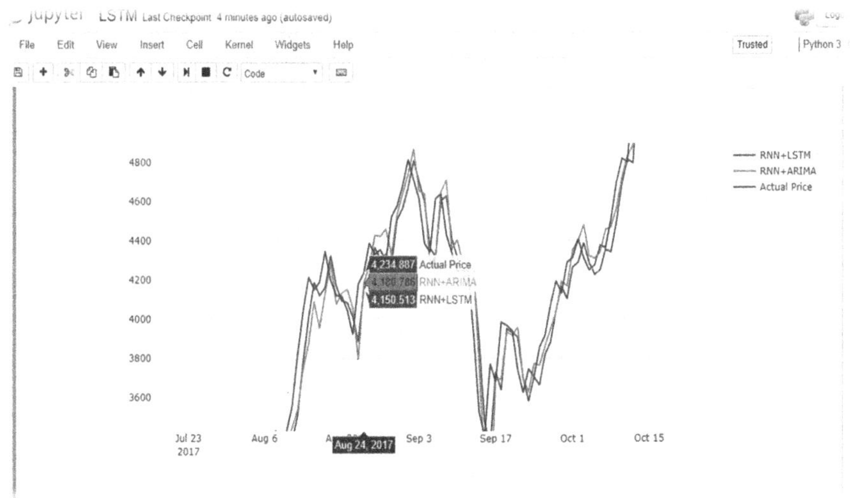

FIGURE 15.1 Bitcoin price prediction on 24 August 2017 – actual price, RNN with ARIMA and RNN with LSTM.

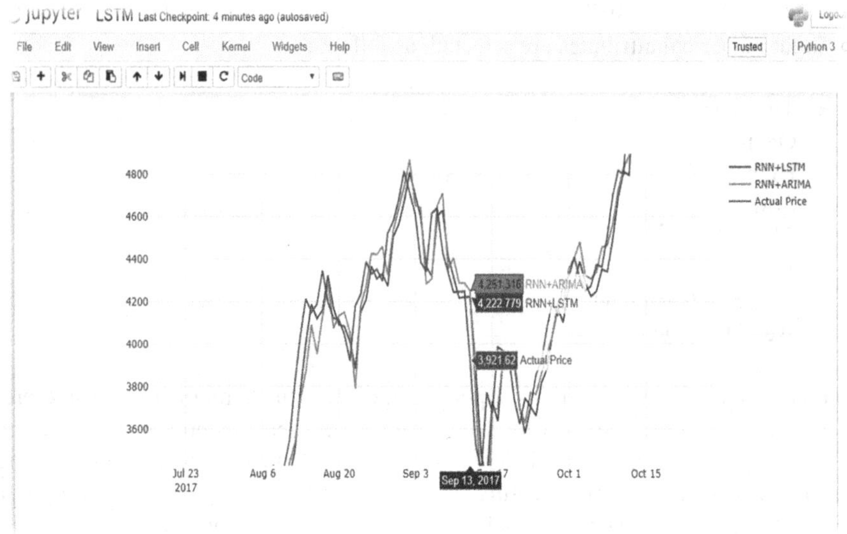

FIGURE 15.2 Bitcoin price prediction on 13 September 2017 – actual price, RNN with ARIMA and RNN with LSTM.

Similarly, an experiment was conducted to predict the bitcoin price on different dates: 11 October 2017, 13 July 2017 and 24 July 2017. The different graphs obtained using RNN with ARIMA and RNN with LSTM indicate that the variation in the bitcoin price is not uniform The experiment results also shows that with an unexpected deep and steep fall in the prices, almost all the time RNN and ARIMA stayed close to the real value, giving a better prediction, which is more reliable than RNN and LSTM. Though the RNN and LSTM performs well, it does not match the actual price of the bitcoin. The improvements of accuracy and efficiency over actual and other selected models are described in Table 15.1 for the year 2016.

The average difference and efficiency of the selected models are computed as follows from Table 15.1.

The Average Difference for the Year 2016:

For RNN+LSTM Model : $ 29.52634
For RNN+ARIMA Model : $ 4.84223

The Efficiency Produced by the Model for the Year 2016:

For RNN+LSTM Model : 95.73135%
For RNN+ARIMA Model : 98.32289%
Overall Improvement : 2.591542%

Similar calculations were also carried out for the year 2017, which are described in Table 15.2.

TABLE 15.1
Improvement of Accuracy and Efficiency over Actual and RNN with LSTM and RNN with ARIMA for the Year 2016

Date	Actual (in USD)	RNN+LSTM (in USD)	RNN+ARIMA (in USD)	Difference (RNN+LSTM)	Difference (RNN+ARIMA)	Efficiency (RNN+LSTM)	Efficiency (RNN+ARIMA)
30 Aug 2016	577.61	552.7314	583.5809	–24.8786	5.9709	95.69283773	98.97685137
15 Sep 2016	612.5537	587.3198	620.2215	–25.2339	7.6678	95.88054076	98.76369974
30 Sep 2016	607.4874	581.3899	614.6232	–26.0975	7.1358	95.70402612	98.83899599
15 Oct 2016	643.3345	611.9614	644.8116	–31.3731	1.4771	95.12336118	99.77092534
30 Oct 2016	701.7091	679.363	711.9496	–22.3461	10.2405	96.81547524	98.56162571
15 Nov 2016	712.497	671.8914	707.6179	–40.6056	–4.8791	94.30094443	99.6895105
30 Nov 2016	739.2463	701.3987	737.1479	–37.8476	–2.0984	94.88024492	99.2846647
15 Dec 2016	779.2243	745.2243	783.7993	–34	4.575	95.63668638	99.41630466
30 Dec 2016	952.4766	929.1219	965.9671	–23.3547	13.4905	97.54800275	98.60342034

TABLE 15.2
Calculation of Accuracy and Efficiency over Actual and RNN with LSTM and RNN with ARIMA for the Year 2017

Date	Actual (in USD)	RNN+LSTM (in USD)	RNN+ARIMA (in USD)	Difference (RNN+LSTM)	Difference (RNN+ARIMA)	Efficiency (RNN+LSTM)	Efficiency (RNN+ARIMA)
15 Jun 2017	2327.565	2558.561	2518.853	230.996	−191.288	90.97164383	92.2183741
30 Jun 2017	2509.231	2485.625	2522.392	−23.606	−13.161	99.9497008	99.5245033
15 Jul 2017	2061.633	2181.748	2189.811	120.115	−128.178	94.49455207	94.2173044
30 Jul 2017	2714.08	2674.032	2685.058	−40.048	29.022	98.4976635	98.93068738
15 Aug 2017	4149.146	4186.364	4091.217	37.218	57.929	99.11097076	98.60383317
30 Aug 2017	4576.087	4504.573	4531.19	−71.514	44.897	99.5875867	99.01887792
15 Sep 2017	3403.445	3475.101	3246.805	71.656	156.64	97.93801677	95.39760449
30 Sep 2017	4290.117	4103.075	4170.027	−187.042	120.09	96.5585811	97.20077564
15 Oct 2017	5610.778	5723.933	5840.793	113.155	−230.015	98.02312501	96.0995206

The average difference and efficiency of the selected models are computed as follows from Table 15.2 for the year 2017.

The Average Difference for the Year 2017:

For RNN+LSTM Model	: $ 27.881
For RNN+ARIMA Model	: $ 17.122

The Efficiency Produced by the Model for the Year 2017:

For RNN+LSTM Model	: 97.23687%
For RNN+ARIMA Model	: 98.80128%
Overall Improvement	: 1.56448%

The variation in accuracy obtained for both models is depicted as a graph in Figure 15.3.

From the selected two models, RNN with ARIMA shows better accuracy of 2.078011% compared with RNN with LSTM. The proposed hybrid model RNN with ARIMA is working as expected and producing outputs with improved efficiency of 2.078011% over the previous similarly available model, which is shown in Figure 15.3. Surprisingly, the combination of stochastic and neural networks works better in combination and produces better accuracy.

15.6 CONCLUSION

The main objective of this model is to improve the existing method of predicting the bitcoin price by applying and adding recurrent loops in the activation function inside the RNN model with ARIMA. The advancements in machine learning techniques

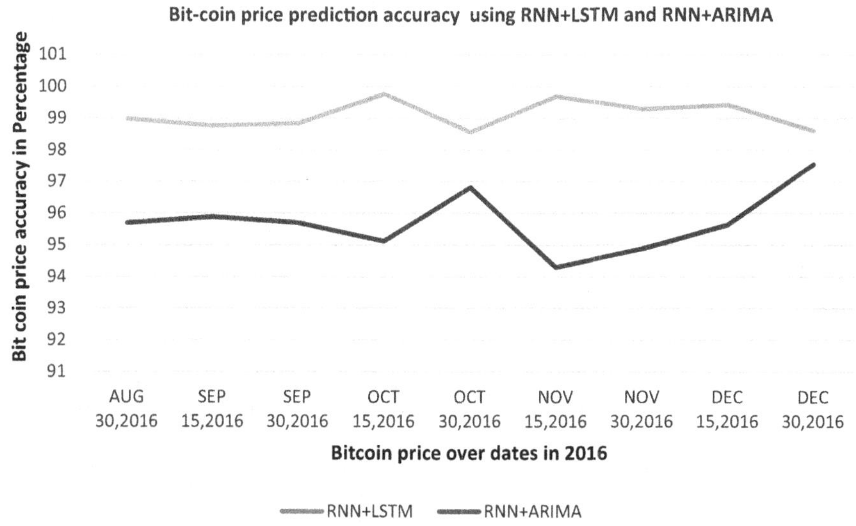

FIGURE 15.3 Bitcoin price prediction accuracy using RNN+LSTM and RNN+ARIMA.

and convolutional neural network able to remember the previous recall of the bitcoin price. This will improve the prediction accuracy of the bitcoin price. Moreover, the difference between predicted accuracy of bitcoin price and actual price of bitcoin will be minimized with deep learning techniques. The proposed methodology should be improved further with advancements in machine learning and deep learning techniques. Moreover, the proposed model considers only quantitative parameters to find the bitcoin price. However, the bitcoin price not only depends on quantitative parameters, and it is also depending on certain number of qualitative parameters (indirect parameters that cannot be measured in units). For example, the qualitative parameters such as global market condition, disease spreading level, pandemic situation due to Covid19, political situation, government authorization for cryptocurrencies, etc. should be considered to predict the bitcoin price.

REFERENCES

Alaraj, M., Majdalawieh, M., & Nizamuddin, N. (2021). Modeling and forecasting of COVID-19 using a hybrid dynamic model based on SEIRD with ARIMA corrections. *Infectious Disease Modelling*, *6*, 98–111.

Awoke, T., Rout, M., Mohanty, L., & Satapathy, S. C. (2021). Bitcoin price prediction and analysis using deep learning models. In *Communication Software and Networks* (pp. 631–640). Singapore: Springer.

Benvenuto, D., Giovanetti, M., Vassallo, L., Angeletti, S., & Ciccozzi, M. (2020). Application of the ARIMA model on the COVID-2019 epidemic dataset. *Data in Brief*, *29*, 105340.

Bousqaoui, H., Slimani, I., & Achchab, S. (2021). Comparative analysis of short-term demand predicting models using ARIMA and deep learning. *International Journal of Electrical & Computer Engineering*, *11*(4), 2088–8708.

Chen, C., Liu, L., & Zhao, N. (2020a). Fear sentiment, uncertainty, and bitcoin price dynamics: The case of COVID-19. *Emerging Markets Finance and Trade*, 56(10), 2298–2309.

Chen, W., Zheng, Z., Ma, M., Wu, J., Zhou, Y., & Yao, J. (2020b). Dependence structure between bitcoin price and its influence factors. *International Journal of Computational Science and Engineering*, *21*(3), 334–345002E

Chen, Z., Li, C., & Sun, W. (2020c). Bitcoin price prediction using machine learning: An approach to sample dimension engineering. *Journal of Computational and Applied Mathematics*, *365*, 112395.

Cretarola, A., Figà-Talamanca, G., & Patacca, M. (2020). Market attention and Bitcoin price modeling: Theory, estimation and option pricing. *Decisions in Economics and Finance*, *43*(1), 187–228.

Dutta, A., Kumar, S., & Basu, M. (2020). A gated recurrent unit approach to bitcoin price prediction. *Journal of Risk and Financial Management*, *13*(2), 23.

Ediger, V. Ş., & Akar, S. (2007). ARIMA forecasting of primary energy demand by fuel in Turkey. *Energy Policy*, *35*(3), 1701–1708.

Felizardo, L., Oliveira, R., Del-Moral-Hernandez, E., & Cozman, F. (2019, October). Comparative study of Bitcoin price prediction using WaveNets, Recurrent Neural Networks and other Machine Learning Methods. In 2019 6th International Conference on Behavioral, Economic and Socio-Cultural Computing (BESC) (pp. 1–6). IEEE.

Guo, H., Zhang, D., Liu, S., Wang, L., & Ding, Y. (2021). Bitcoin price forecasting: A perspective of underlying blockchain transactions. *Decision Support Systems*, *151*, 113650, ISSN 0167-9236, https://doi.org/10.1016/j.dss.2021.113650.

Gupta, A., & Nain, H. (2021). Bitcoin price prediction using time series analysis and machine learning techniques. In *Machine Learning for Predictive Analysis* (pp. 551–560). Singapore: Springer.

Ho, W. T., & Yu, F. W. (2021). Predicting chiller system performance using ARIMA-regression models. *Journal of Building Engineering*, *33*, 101871.

Ibikunle, G., McGroarty, F., & Rzayev, K. (2020). More heat than light: Investor attention and bitcoin price discovery. *International Review of Financial Analysis*, *69*, 101459.

Ismail, M. T., Shah, N. Z. A., & Karim, S. A. A. (2021). Modeling solar radiation in peninsular Malaysia using ARIMA model. In *Clean Energy Opportunities in Tropical Countries* (pp. 53–71). Singapore: Springer.

Jalali, M. F. M., & Heidari, H. (2020). Predicting changes in Bitcoin price using grey system theory. *Financial Innovation*, *6*(1), 1–12.

Ji, S., Kim, J., & Im, H. (2019). A comparative study of bitcoin price prediction using deep learning. *Mathematics*, 7(10), 898.

Kaggle. (2021, May 01). Bitcoin data at 1-min intervals from select exchanges, Jan 2012 to March 2021. retrieved on 5, May' 2021 https://www.kaggle.com/mczielinski/bitcoin-historical-data/data

Kavasseri, R. G., & Seetharaman, K. (2009). Day-ahead wind speed forecasting using f-ARIMA models. *Renewable Energy*, *34*(5), 1388–1393.

Khedr, A., Arif, I., Pavijaraj, P. V., El-Bannany, M., Alhashmi, S. S. M. (2021). Cryptocurrency price prediction using traditional statistical and machine learning techniques: A survey. *Intelligent Systems* in Accounting, *Finance* and *Management*, *28*. 3–34. https://doi.org/10.1002/isaf.1488

Kinasz, W. (2021). Use of artificial neural networks and the ARIMA model for short-term stock indices forecasts (Doctoral dissertation, Instytut Elektroenergetyki).

Liu, M. D., Ding, L., & Bai, Y. L. (2021). Application of hybrid model based on empirical mode decomposition, novel recurrent neural networks and the ARIMA to wind speed prediction. *Energy Conversion and Management*, *233*, 113917.

Livieris, I. E., Kiriakidou, N., Stavroyiannis, S., & Pintelas, P. (2021). An advanced CNN-LSTM model for cryptocurrency forecasting. *Electronics*, *10*(3), 287.

Lv, T., Wu, Y., & Zhang, L. (2021, April). A traffic interval prediction method based on ARIMA. *Journal of Physics: Conference Series*, *1880*(1), 012031). IOP Publishing.

Patel, M. M., Tanwar, S., Gupta, R., & Kumar, N. (2020). A deep learning-based cryptocurrency price prediction scheme for financial institutions. *Journal of Information Security and Applications*, *55*, 102583, ISSN 2214-2126, https://doi.org/10.1016/j.jisa.2020.102583.

Pintelas E., Livieris I.E., Stavroyiannis S., Kotsilieris T., Pintelas P. (2020) Investigating the problem of cryptocurrency price prediction: A deep learning approach. In Maglogiannis I., Iliadis L., Pimenidis E. (eds.), *Artificial Intelligence Applications and Innovations. AIAI 2020. IFIP Advances in Information and Communication Technology* (vol. 584). Cham: Springer. https://doi.org/10.1007/978-3-030-49186-4_9

Poongodi, M., Vijayakumar, V., & Chilamkurti, N. (2020). Bitcoin price prediction using ARIMA model. *International Journal of Internet Technology and Secured Transactions*, *10*(4), 396–406.

Rathan, K., Sai, S. V., & Manikanta, T. S. (2019, April). Crypto-currency price prediction using decision tree and regression techniques. In 2019 3rd International Conference on Trends in Electronics and Informatics (ICOEI) (pp. 190–194). IEEE.

Shankhdhar, A., Singh, A. K., Naugraiya, S., & Saini, P. K. (2021, April). Bitcoin price alert and prediction system using various models. *IOP Conference Series: Materials Science and Engineering*, *1131*(1), 012009. IOP Publishing.

Shin, M., Mohaisen, D., & Kim, J. (2021, January). Bitcoin price forecasting via ensemble-based LSTM deep learning networks. In 2021 International Conference on Information Networking (ICOIN) (pp. 603–608). IEEE.

Sin, E., & Wang, L. (2017, July). Bitcoin price prediction using ensembles of neural networks. In 2017 13th International conference on natural computation, fuzzy systems and knowledge discovery (ICNC-FSKD) (pp. 666–671). IEEE.

Velankar, S., Valecha, S., & Maji, S. (2018, February). Bitcoin price prediction using machine learning. In 2018 20th International Conference on Advanced Communication Technology (ICACT) (pp. 144–147). IEEE.

Yang, H., Li, X., Qiang, W., Zhao, Y., Zhang, W., & Tang, C. (2021). A network traffic forecasting method based on SA optimized ARIMA-BP neural network. *Computer Networks*, *193*(3). 108102. 10.1016/j.comnet.2021.108102108102.

16 Blockchain versus IOTA Tangle for Internet of Things

The Best Architecture

C. P. Igiri, Deepshikha Bhargava, C. Udanor and A. R. Sowah

CONTENTS

DOI: 10.1201/9781003138082-16

16.1 INTRODUCTION TO DISTRIBUTED LEDGER TECHNOLOGY AND INTERNET OF THINGS

This section gives a brief introduction to two distributed ledger technologies (DLTs), blockchain technology and IOTA Tangle. A DLT is a digital system that keeps track of transactions and assets and simultaneously replicates the record at multiple locations. Traditional blockchain technology and IOTA adopt the principle of DLT with architectural disparity. A brief overview of the Internet of Things (IoT) is also given here. It is all about the distinguishing features of the two DLTs mentioned here and their relation to IoT applications.

16.1.1 Introduction to Blockchain Technology

Blockchain technology is a DLT with highly secured features. It is immutable, transparent and decentralized, with cryptographic hashing. This significant characteristic of blockchain technology enlists it as a solution to the cost-intensive, insecure and inefficient nature of modern business platforms (Bashir, 2018). It consists of digital information organized as "blocks" and recorded in public databases known as "chains". Blockchain technology forms the bedrock of the Bitcoin network. Haber and Stornetta developed blockchain technology in 1991 to create a time-stamped and tamperproof document (Haber and Stornetta, 1991). However, the technology gained popularity in 2009 through Satoshi Nakamoto as the backbone of Bitcoin technology. Aside from the financial application, blockchain technology has been found useful in many sectors, including healthcare, industry, supply chain and many more. Deloitte's 2020 study on 1400 companies across 14 regions showed that 82% of the companies would be hiring blockchain professionals in the next year (Pawczuk et al., 2020).

Blockchain technology consists of three parts. First, the transaction information includes the date, time and type. The second component is the information about the initiator of the transaction (digital signature). Then the unique code identity of every block is referred to as a cryptography hash. In other words, a block consists of a transaction, a digital signature and a hash function. As soon a transaction is initialized, it goes through a vetting and verification process on a network of authenticated computer nodes known as miners. Miners are computer networks that validate and approve a new block by solving a puzzle using a cryptography hash algorithm. The verified transaction is approved, and a hash function is generated before recording it on the chain. It is replicated on the decentralized network for public view.

Blockchain technology, like any other technology, has merits and demerits. Among the several advantages of blockchain technology, one is transaction record accuracy. It does not require a third party during the transaction verification processes. Transaction verification requires thousands of authenticated network nodes; 51% of the nodes must approve a transaction before it is accepted as a valid transaction, thereby reducing or possibly eliminating the risk of error. Second, it is relatively cost-effective. The absence of a third party, such as a bank or government organization, eradicates these authorities' transaction costs. Additionally, as it is a decentralized system, information recorded in the ledgers is tamperproof. Data cannot be modified without the approval of 51% of the verification nodes (computer network or miners). It is noteworthy to mention that any transaction appended to the network cannot be retrieved or edited. Another advantage of blockchain technology is that it ensures transaction efficiency. A centralized network such as a bank works only 5 days a week within business hours. Blockchain technology, on the other hand, works 24/7. This implies that transactions are initiated and completed within a few hours in a blockchain network. It is especially beneficial for cross-border trade, where time zone variation could delay payment confirmation processes. Other merits of blockchain include transparency, privacy, security and many more.

Despite the vast benefits of blockchain technology, it also has some drawbacks. The technology itself incurs high computation power, which is associated with cost. This cost stems from the hardware cost of the "proof of work" (PoW) concept employed to validate a transaction. It is, however, more cost-effective for the user than the miners. Secondly, the transaction speed is relatively slow. Bitcoin, for example, requires about ten transactions per block. This speed is slow compared with other financial processes, such as the legacy brand Visa, which completes 24,000 per second.

16.1.2 Introduction to IOTA Technology

According to the IOTA Foundation (Robert Shorten, Bill Buchanan, Serguei Popov, n.d.), IOTA is an open-source DLT, which enables devices connected in the Tangle to transact micropayments and exchange immutable data. The Tangle is a DLT that employs a mathematical concept known as directed acyclic graph (DAG). Unlike other DLT networks, IOTA removes the miners without sacrificing decentralization. The rationale behind the feeless feature of the Tangle network is the absence of miners. To achieve this, everyone in the network participates in validating transactions. The two core distinguishing features of the IOTA are microtransactions and speed. First, microtransactions are transactions that could be done for a fraction of a cent. Most cryptocurrencies do not support micropayments, like the purchase of coffee, due to high network fees for miners. Second, the IOTA network speed is breakneck. Usually, in cryptocurrency, speed is measured in transactions per second (TPS). The speed in the IOTA Tangle is scaled by the number of contributors in the network. The more people participating in the network, the faster the speed, thereby alleviating the drawbacks that other cryptocurrencies like Bitcoin face today. These features inspire the IOTA purpose, supporting and facilitating the machine-to-machine (M2M) economy.

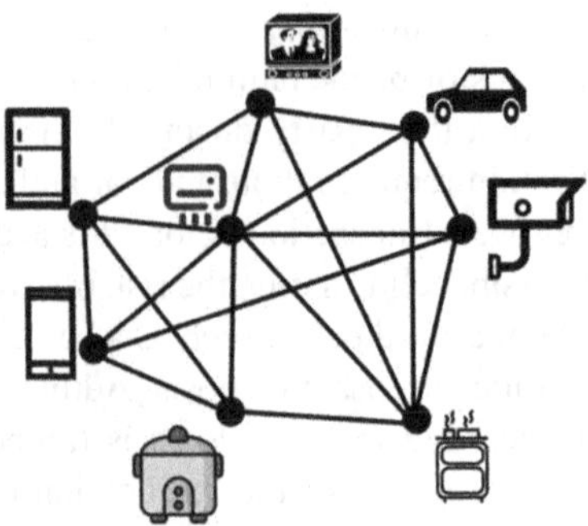

FIGURE 16.1 Machine2Machine communication

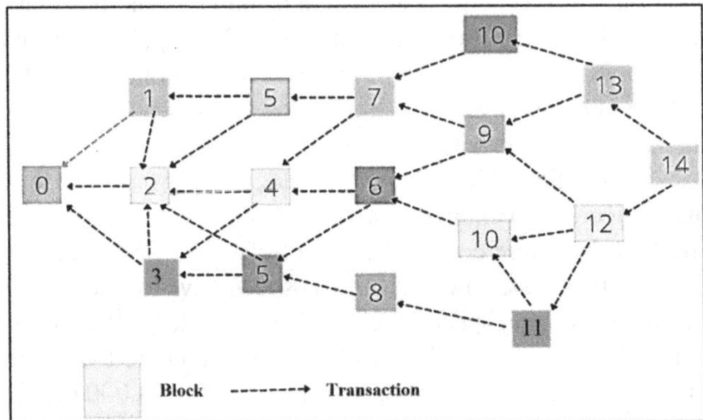

FIGURE 16.2 Transaction blocks in IOTA Tangle. (From *An illustration of transaction blocks in IOTA Tangle*. (2018, September 27). [Image]. https://www.Blockchainguide.Biz/Tag/Iota/.)

IOTA is a protocol for IoT; unlike blockchain, it is highly scalable. Three components constitute the IOTA architecture: the node, the Tangle and the client.

- The node is the network of devices that validates transactions to ensure the Tangle's integrity.
- The clients are network users.
- The Tangle is the network of validated transactions simulated across all the nodes in the network. Every valid transaction is appended to two previous ones directly or indirectly to form a DAG. For example, in Figure 16.1, Block 3 is directly attached to block number 1 but indirectly connected to the genesis block 0.

16.1.3 Introduction to Internet of Things

IoT is a technology that brings together various "things" such as sensors, embedded devices, controllers, software, etc. together under one physical network. Things could be objects of the physical world, including cameras, robotic actuator arms, light bulbs, fans, air conditioners, etc. Or, they could be of the information or virtual world, such as network packets, bits of data, biometric data and so on. These things are recognized and integrated into a communication network with their unique identifiers. The associated information possessed by these things could be static or dynamic.

The integration of sensor and radiofrequency technologies makes IoT a distributed network based on the omnipresent Internet hardware resources (Han et al., 2012). IoT as an emerging technology comes on the heels of such computing fields as communications networks and global roaming technologies, involving the addition of sophisticated technologies such as remote communication, remote data transmission technologies, sea and earth measures information, intelligent data analytics, control technologies and DLTs.

The evolution of IoT is due to the convergence of various technologies in place today. Such technologies include wireless communication networks such as the 4G and 5Gs, satellite communications systems, wireless sensor networks (WSNs), embedded systems, offices, smart home automation systems, commodity sensors, etc. More recently, machine learning, DLTs (blockchain, IOTA) and real-time data analytics have come to play significant roles in the IoT ecosystem (Gupta et. al, 2020; Petiwala et. al., 2021). The International Telecommunication Union (ITU), which started a global standard initiative on IoT (IoT-GSI), defined IoT well in 2015 in its Reference ITU-T Y.2060 as "a universal structure for the information systems, permitting progressive facilities by interrelating both physical as well as virtual things on the basis of prevailing and growing interoperable informative and communication technologies" (Dabeesing, 2017). This standard recommendation implies that through identification, data capturing and processing capabilities, IoT should be able to make full utilization of things to provide every kind of service to users for confirming security. IoT technologies enable objects prepared with sensors, actuators, processors and transceivers to communicate. However, security and privacy are still a concern to the IoT ecosystem, and this is where DLTs' role is critical.

16.2 DESCRIPTION OF BLOCKCHAIN AND IOTA ARCHITECTURE

The two distinguishing features of the IOTA Tangle from the blockchain are: first, the IOTA is fee free, while blockchain attracts network charges; second, the blockchain employs miners, while IOTA does not. These two DLTs use different data structures and consensus protocols. A brief description of the data structures and consensus protocol is necessary to drive home the distinguishing features.

16.2.1 Data Structure

The data structure is how data and assets are stored in the computer for efficient manipulation. Blockchain employs a hash-graph sequential data structure to attach transactions to a chain. When the miners validate a transaction, only one block is added to the chain at a time; see Figure 16.2. Consequently, this results in a blockchain bottleneck, which is the reason for the slow TPS in many cryptocurrencies like Bitcoin today.

On the contrary, the Tangle employs the DAG data structure to append validated blocks to the Tangle. A transaction is attached to two previous blocks. Multiple blocks could be included to the Tangle simultaneously, as shown in Figure 16.4. The numerous blocks attached simultaneously eliminate the bottleneck in blockchain architecture, as shown in Figures 16.2 and 16.3.

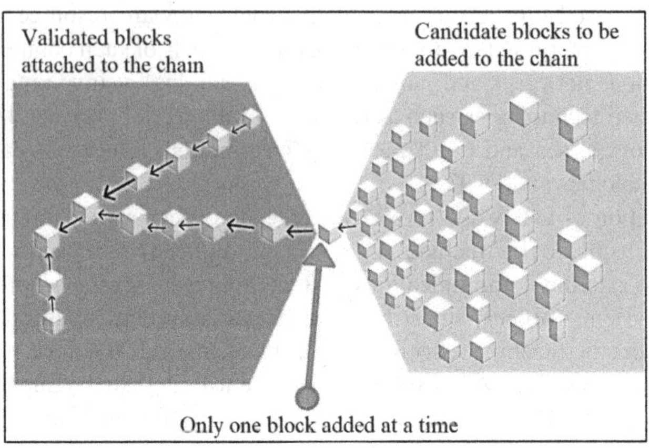

FIGURE 16.3 Blockchain data structure.

16.2.2 Consensus Protocol

In blockchain, miners and users constitute the network. The miners validate the transactions for integrity using the PoW protocol, while the users initiate transactions. The mining incurs enormous computing power. The users provide an incentive to the miners to execute this function, thereby imposing a high network fee on the blockchain network. A valid transaction is referred to as a block. Blockchain adopts the PoW to validate and secure the network from attackers.

The IOTA Tangle, on the other hand, does not require miners to validate transactions. It preferably employs "coordinators" temporarily for transaction validation and confirmation. The PoW is used just to discourage spam transactions.

Note that the IOTA Foundation declares that the use of "coordinators" is temporary and should be replaced for better performance (Robert Shorten, Bill Buchanan, Serguei Popov, n.d.).

16.2.3 IoT Architecture and Connectivity

The development of ever-present computing infrastructure in which a vast array of heterogeneous digital devices that are uniquely identified and interconnected for interacting with one another in real time, collecting and transmitting data, requires an amalgamation of original and valuable skills. It is only possible through the addition of these technologies. Figure 16.4 depicts these connectivity technologies, showing the signal range of each solution.

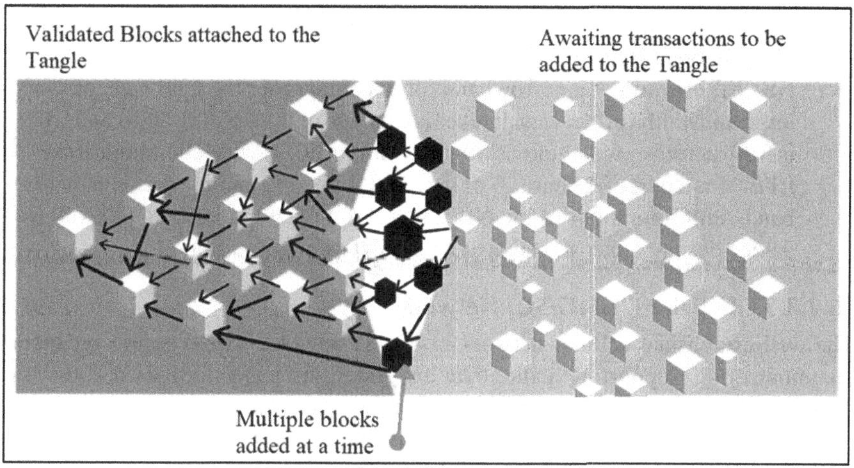

FIGURE 16.4 IOTA Tangle data structure.

16.2.3.1 Radio Frequency Identification (RFID)

It would be complicated to keep track of the various sensors in a network of so many sensors. RFID provides the technology that makes objects uniquely identifiable in such a network. Its small size and low cost make it possible for it to be easily integrated into an entity (Bansal and Rana, 2017). It possesses a transceiver microchip, which may appear as a sticker for use as object tags. RFID can be active or passive. Unlike active tags, which have batteries, passive tags are triggered when activated by some stimulus. The frequency range of RFID is divided into four bands depending on the application. These include: (1) low frequency (135 kHz or less); (2) high frequency (13.56 MHz); (3) ultra high frequency (862 to 928 MHz); and (4) microwave frequency (2.4G, 5.80).

16.2.3.2 Low Power Wide Area Networks (LPWANs)

These are the new technologies that were developed due to the demand for low-power and low-cost wireless communications suitable for IoT. LPWANs provide long-range signals on small, low-cost devices with batteries that are built to last for many years. These include Sigfox, LoRa, NB IoT and LTE-M, LPWANs can connect to any type of sensor and find their applications in the areas of environmental monitoring, consumables monitoring, asset tracking, etc.

- SigFox, an open standard, is most suitable for applications in the lowest bandwidth region. In Europe and the US, it operates in the 868 MHz and 900 MHz frequency bands, respectively. SigFox is also recommended for projects that require the lowest energy consumption.
- LoRa was developed by Semtech, a chip manufacturing company, and offers much higher bandwidth than other LPWAN technologies. It also provides an open standard.
- NB IoT, operating on narrowband, offers an appropriate LTE category for low-bandwidth IoT devices. It takes advantage of the existing GSM and LTE infrastructures to facilitate low bandwidth for IoT device communication.
- LTE-M is a part of Release 13 of the 3GPP standard, an improvement on the bandwidth limitations of the NB IoT; it is aimed at reducing complexity in device design and cost, and also lowering power consumption.

16.2.3.3 Cellular (3G/4G/5G) Networks

The well-established cellular network offers reliable, stable and secure broadband communication, supporting audio, data and video. Its major limitation is the high operational cost. While these networks may not be viable for IoT due to power requirements, they fit better into applications like associated fleet management and transport logistics.

16.2.3.4 ZigBee

By the IEEE 802.15.4 standard, ZigBee is a short-range, low-powered wireless IoT technology generally organized in a mesh topology for extending the area of exposure by relaying data over multiple sensor nodes. ZigBee provides higher data rates when compared with LPWAN, but the mesh configuration poses a limitation to power-efficiency. Due to its short physical range of less than 100 m, ZigBee is used for nearby applications, such as in-home automation, security and energy management.

16.2.3.5 WiFi

It's often used as a solution for large networks of battery-operated IoT sensors. It can be used in industry, schools or homes. It finds application in a vast number of domains, including smart buildings, home gadgets, security cameras, etc. The following WiFi frequencies are used by IoT: 169, 433, 863–870, 2400 to 2483.5, 5150–5350 and 5470–5716 MHz (Zhao and Johnson, 2021).

16.2.3.6 Bluetooth

As a short-range technology, it finds application in the personal area network. It is used in wearable devices like health and fitness monitoring devices. BLE-enabled devices are generally utilized in combination with electronic devices, mainly smartphones that help as a centre for transmitting data to the cloud.

Behr et al. (2020) provide a graphic summary of the application areas of the various IoT connectivity technologies in Figure 16.5.

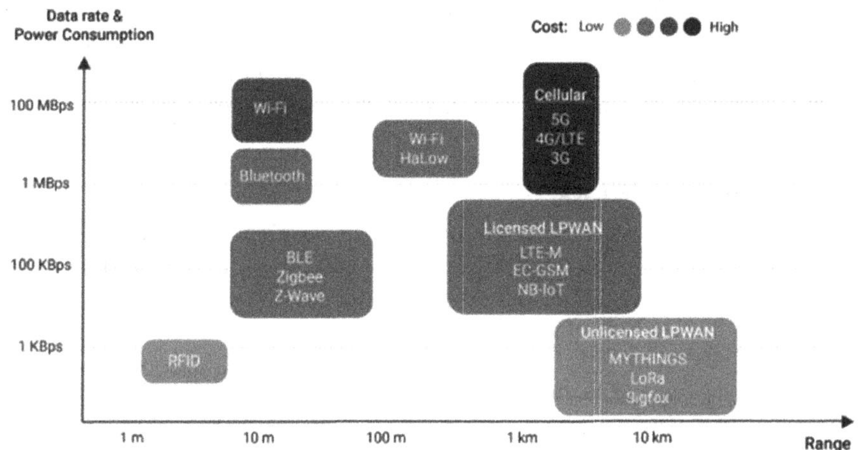

FIGURE 16.5 IoT connectivity technologies and solutions. (From Behr, A., et al., *Behr Technologies*, 2020, https://behrtech.com/solutions/)

16.2.4 Design and Operational Architecture

The domain of IoT encompasses a plethora of technologies, making it difficult to use a single architecture to model IoT implementation. Therefore, it is most likely that a number of reference models will coexist in the design of a reference model for IoT (Jamali et al., 2020). For this reason, IoT should be open enough to provide open protocols for supporting current network applications. Sethi and Sarangi (2017) argue that there is no agreement on a single universally accepted framework, but different authors have proposed different architectures. Two such architectures are the three-layer and five-layer architectures, as shown in Figure 16.6.

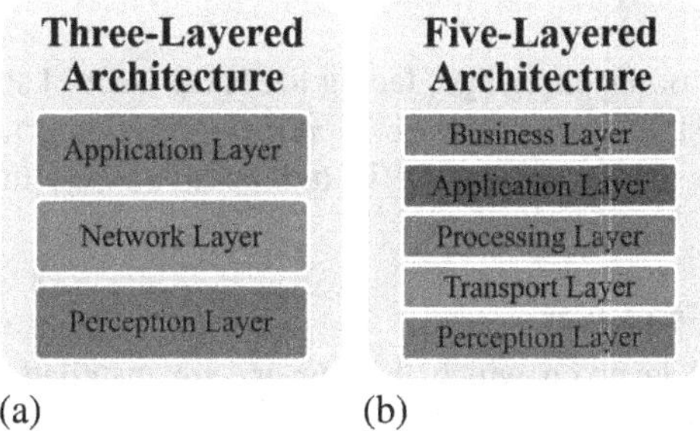

FIGURE 16.6 (a) Three-layered and (b) five-layered architecture of IoT. (From Sethi, P. and Sarangi, S. R., *Journal of Electrical and Computer Engineering*, 9324035, 25, 2017.)

The IoT architecture in Figure 16.6a has a basic high-level three-tier structure, while Figure 16.6b has three additional layers to make up a five-tier structure. These architectures are classified on the basis of their protocols. These will be briefly explained. The layers common to both architectures perform the same functions.

16.2.4.1 Application Layer

In both architectures, this layer is responsible for defining and delivering various application areas where IoT can be used, such as e-health, air or water quality monitoring, etc.

16.2.4.2 Network Layer

This defines the interconnectivity of things and various media through which sensor data can be transmitted, such as 4G, 5G, RFID, WiFi, WiMaX, ZigBee, GSM, 3G, etc., with protocols like IPv4, IPv6, MQTT, DDS, etc.

There are five key aspects to consider in network connectivity (Konezny et al., 2021):

- Range: How close or how far are the devices going to be to each other? Is the deployment on a single floor or city-wide?
- Data Rate: What amount of bandwidth will be required, and how often does the data change? The key here is to know the type of data beforehand. Will the network transmit text, audio or video?
- Power: Though most IoT devices are low-powered, it is essential to know the power source: mains or battery power supply? And, for how long will these devices be expected to be on? It is also essential here to plan for power management schemes to conserve power, especially if the devices are installed in remote locations.
- Frequency: Each device has an operational frequency range specified by the manufacturer according to industry standards. But, there is the need to consider channel blocking and signal interference during design and deployment.
- Security: One of the challenges facing IoT is the issue of security. Knowing beforehand if the sensors will be supporting mission-critical applications is vital. Studies show that the DLT attempts to address the security issues in IoT.

16.2.4.3 Perception Layer

This is the physical layer on which the sensors are installed for gathering information from their environments. Environmental parameters like temperature, relative humidity, water level, etc., are sensed and collected by smart devices and then transmitted.

16.2.4.4 Business Layer

This is specific to the architecture in Figure 16.7. This layer does more housekeeping in the IoT system, where it manages the user's application and data privacy, business and profit models, and so on.

16.2.4.5 Processing Layer

This is the middleware layer. The processing includes data storage functions employing the use of databases and data analytics using technologies like edge, fog and cloud computing, big data, etc. Third-party vendors frequently provide this layer's functions.

16.2.4.6 Transport Layer

The role of this layer, as the name implies, is to transfer data from the sensors at the perception layer to be processed at the processing layer using network layer functions.

Bansal and Rana (2017) present a six-layer architecture similar to Figure 16.7. It only differs in the inclusion of a coding layer, which they say provides identification of objects of interest. The layer assigns unique IDs to each object to make it identifiable in the network.

16.2.4.7 Device Capability Exposure (DCE)

This is defined as a functional entity to manage the connected IoT devices in an IoT area network, to expose the capabilities of the connected IoT devices to IoT applications, and to support the IoT applications to access the exposed device capabilities (Zhao and Johnson, 2021). This reference model is shown in Figure 16.7.

Key IoT Verticals	LPWAN (Star)	Cellular (Star)	Zigbee (Mostly Mesh)	BLE (Star & Mesh)	Wi-Fi (Star & Mesh)	RFID (Point-to-point)
Industrial IoT	●	○	○			
Smart Meter	●					
Smart City	●					
Smart Building	●		○	○		
Smart Home			●	●	●	
Wearables	○			●		
Connected Car					○	
Connected Health		●		●		
Smart Retail		○		●	○	●
Logistics & Asset Tracking	○	●				●
Smart Agriculture	●					

● Highly applicable ○ Moderately applicable

FIGURE 16.7 Summary of IoT applications connectivity solutions. (From Behr, A., et al., *Behr Technologies*, 2020, https://behrtech.com/solutions/)

16.3 REVIEW OF RELATED WORKS

This section reviews literature related to blockchain, IOTA Tangle and IoT.

16.3.1 Related Work in Blockchain Technology

Studies show that blockchain application is progressively extending from cryptocurrency to other industrial domains, including health (Engelhardt, 2017; Radanovic and Likic, 2018); energy grid and smart cities (Sun et al., 2016; Gupta and Vyas, 2021); banking and finance (Hyvarinen et al., 2017); business economics (Kim and Laskowski, 2018); and document protection (Haber and Stornetta, 1991; O'Dair and Beaven, 2017). A systematic study by Xu et al. (2019) shows that Bitcoin takes 80% of the blockchain research share, while the remaining 20% goes to other domains. This section reviews blockchain applications in IoT and smart grids.

IoT has broad applications, typically in healthcare devices, energy grids, and lots more. The machines are heterogeneous and lack standardization. Among other things, these generic features expose the gadgets to security and privacy risks (Dorri et al., 2017).

The PoW model used in verifying transactions for security purposes in the conventional blockchain consumes high computation resources and is not ideal for IoT implementation. It results in traffic and processing overheads (Dorri et al., 2017). Therefore, adopting the traditional blockchain in IoT might solve the insecurity and privacy problem but might compromise the response time. The response time is critical, especially in a heterogeneous environment like IoT. Dorri et al. (2017) proposed a modified blockchain technology called overlay blockchain to address this challenge. The result showed an improved performance compared with the traditional blockchain technology (Dorri et al., 2017).

Novo (2018) proposed another improved blockchain technology for access control in IoT. The author adopts the management hub to decentralize access control among IoT devices using blockchain technology. Typically, access control in IoT devices is centralized. Therefore, one point of failure could be catastrophic. As the number of devices increases, the network becomes more insecure. Novo's research focus was to address scalability issues with an extensive network of numerous IoT devices in real time. Although the management hub provided a relatively better solution than other access control solutions like DOAuth and "fairAccess", process overhead remains a significant challenge (Novo, 2018).

The smart grid is the integration of electric power producers and consumers. It is a concept that enables small-scale power generators to produce and sell energy to consumers. Anak Agung GdeAgung, Rini Handayani (in press) proposed a smart grid based on blockchain technology. The research focuses on electricity generation and supply between producers and consumers to ensure transparency, security, immutability and efficiency. The system is implemented on the Ethereum platform; therefore, the transaction token is Ether. Ethereum is a type of blockchain network that majors in developing smart contracts. Ethereum uses PoW as a consensus mechanism. According to the authors, although the system attempts to address transparency and immutability, it has some limitations. Only 15 transactions/second could be processed on the Ethereum platform, resulting in a slow response time. Also, the cost of sustaining miners used in PoW would increase the energy consumption cost.

16.3.2 Related Works on IOTA

As said earlier, IOTA Tangle utilizes a data structure specifically designed for IoT. It is fee free and validates transactions without miners, unlike blockchain architecture. Blockchain employs the PoW concept as a consensus mechanism to validate transactions. However, Vigneri and Welz (2020) designed an adaptive rate control protocol using PoW to prevent spam transactions among IoT devices. Cullen et al. (2021) proposed a new architecture DLT that supports "reputation based on Sybil protection" to be integrated into the IOTA Tangle (that is, DAG). The authors' (Cullen et al., 2021) goal is to address the access control problem and the computation limitation of devices. The "reputation-based Sybil protection" is used to substitute for PoW, used in traditional blockchain architecture.

16.3.3 IoT Background and Related Works

Kevin Ashton first coined the term *Internet of Things* in 1999 during his proposal of integrating RFID into Procter and Gamble's supply chain (Ashton, 2009). His idea was that since people are usually very busy, there was a need to use RFID sensors to empower computers to gather information randomly by themselves without being limited by humans entering the data. IoT is seen as the Internet's future, drastically reducing a human-to-human interaction while increasing M2M transactions. It promises to unify everything in our world under one architecture while at the same time giving us control over many things and keeping us informed on the goings-on around us (Bansal and Rana, 2017). IoT as a new revolution of the Internet describes a future with the possibility of connecting all physical devices (Yehia et al., 2015), which will communicate among themselves independently of human intervention. These devices will affect all facets of our everyday life, such as in monitoring our health status, our homes and offices, and water and air quality, among others. The history of IoT can be traced back to the early telemetry system, which began in Chicago around 1912, in which telephone lines were used to monitor data from power plants (Zennaro, 2016). In the 1930s, telemetry expanded to weather monitoring using devices known as radiosondes. The Sputnik, launched by the Soviet Union in 1957 during the space race era, became the basis for aerospace telemetry, which later gave birth to today's global satellite communications. Also, according to Zennaro (2016), M2M technologies began in the 1980s as wired communication began to advance towards wireless in the 1990s. Several enabling technologies that have aided the rise of IoT include ubiquitous connectivity, widespread adoption and expansion of the IP address regime, computing economics, miniaturization, advances in data analytics and the rise of cloud computing.

Fundamental characteristics of IoT include interconnectivity, heterogeneity, dynamism and enormity. Regarding interconnectivity, almost anything can be interconnected to this global information and communications infrastructure. These devices are mostly heterogeneous, coming from different hardware manufacturers and network architectures. Despite this heterogeneity, the devices can integrate and interface seamlessly with one another based on IoT standards. IoT devices are dynamic in their ability to change state quickly in the shortest possible time. They could move from sleep/standby to the awake or active state. They could promptly activate various sensory modes as soon as they detect a state change, moving from the disconnected to the connected state, for example. The enormous scale of IoT interconnectivity would be at least an order of magnitude exceeding the devices connected to the current Internet (Zennaro, 2016). Cisco predicted that by 2020, more than 25 billion things will be interconnected on the IoT network globally (Bansal and Rana, 2017). This magnitude far exceeds what the current TCP/IP networks can handle without compromising security and quality of service (QoS) standards. The IoT promises to be an open architecture.

16.4 INNOVATIVE FEATURES OF IOTA FOR THE INTERNET OF THINGS

IoT is a data-sharing network whereby interconnected devices can exchange data among themselves. Specifically, IOTA provides a platform that enables IoT devices to build applications that could run on the Tangle. Some IoT devices have sufficient inherent power to compute the IOTA Tangle's PoW requirement, while others do not. IOTA provides optional middleware to aid power-deficient IoT devices to calculate PoW. Besides, IOTA has various hardware that could support a variety of applications. Typically, CryptoCore can support applications that require fast, dedicated PoW and secure memory (Foundation-IOTA, 2021). Others that could support WiFi and low Bluetooth energy are also available. Further, single-board computers for building highly power-consuming applications are publicly available. Examples of boards supported by IOTA are nRF52 and ESP32 single-board computers, among others IoT DCE, as shown in Figure 16.8 (Foundation-IOTA, 2021).

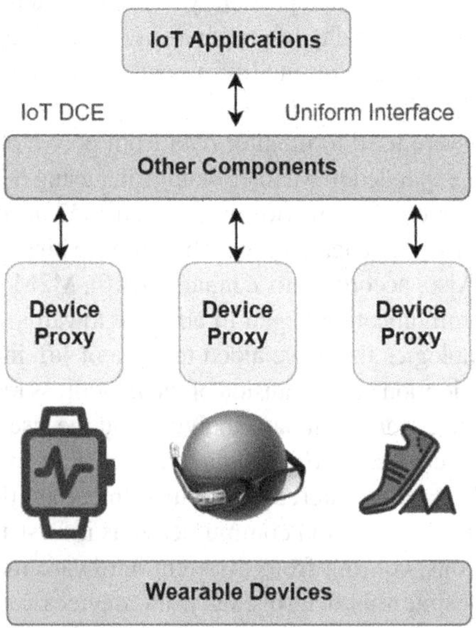

FIGURE 16.8 Overview of IoT DCE. (From Zhao, H. and Johnson, M., *Global Information Infrastructure, Internet Protocol Aspects, Next-generation Networks, Internet of Things, and Smart Cities*, International Telecommunication Union, 2015.)

16.4.1 IOTA Projects

The IOTA Foundation has developed massive technological backbones for building a variety of products across various industries. For instance, in the automobile industry, there is a digital wallet framework that would provide access for cars to autonomously pay for services like battery charging, parking, toll fees and lots more. The Digital Twins is another IOTA framework that provides transparent tracking for the lifecycle of the vehicle. It could be used to track vehicle ownership, usage, fraud detection, etc.

Currently, there are numerous real-world products already running on the platforms mentioned earlier. For brevity, this chapter cites two existing IOTA projects. First, there is a Jaguar Land Rover car wallet that runs on the Tangle. It enables automated payment services, such as sending and receiving payments (IOTA-Foundation, n.d). Second is smart charging with IOTA ElaaNL (IOTA-Foundation, n.d). This is a plug and play device based on the Tangle that enables electric vehicles to pay automatically for charging a car. Moreover, IOTA provides opportunities in various application domains, including healthcare, digital entities, smart cities and supply chain, to name a few. All in all, IOTA has a large market share in the 4.0 industry revolution.

16.5 HOW DOES THE ARCHITECTURE OF BLOCKCHAIN AND IOTA AFFECT THE INTERNET OF THINGS?

This section discusses the distinction between the two DLTs concerned, blockchain and IOTA, and how it affects efficiency in IoT networks. Specifically, two features will be considered in this comparison: Consensus protocol and data structure.

16.5.1 Consensus Protocol

Various DLTs adopt various consensus protocols to secure the network. Some are implemented in public DLTs, while others are used in private ones. The popular ones are PoW, proof of stake (PoS), Byzantine Agreement method, Raft, VRF-Based method, Sharding-based method and their respective variants. Generally, most DLTs, including Ethereum, Cardano and Binance, adopt the consensus as discussed in algorithms for transaction verification (Salimitari and Chatterjee, 2018). Unfortunately, they involve high computing overhead, high network overhead, high storage overhead, low throughput and many more (Salimitari and Chatterjee, 2018). These limitations make them unsuitable for low-slung resource-constrained IoT devices. On the other hand, Practical Byzantine Fault Tolerance (PBFT), a variant of the Byzantine Agreement method adopted in the hyper ledger, does not involve high computing power, low computing overhead, low network, and inadequate storage overhead. However, the hyper ledger is a private blockchain. Therefore, its network capacity is limited (accommodating only a few devices) (Salimitari and Chatterjee, 2018). Readers may refer to Salimitari and Chatterjee (2018) for more details on diverse consensus protocols.

Conversely, IOTA implements the Markov Chain Monte Carlo (MCMC) and a tiny PoW to verify and decide valid blocks (Kljajic, 2018). In this algorithm, every node in the network must approve two previous transactions directly or indirectly to ensure that the entire network achieves consensus (Foundation-IOTA, 2021). This method involves little computing, storage, and network overhead, provides high throughput and is highly scalable. An increase in transaction traffic proportionally speeds up the network. As per IOTA Foundation, this node is temporary and referred to coordinators (Foundation-IOTA, 2021). This feature is most appropriate for IoT devices with low resource constraints. IOTA application hardware are shown in Figure 16.9.

FIGURE 16.9 Sample hardware supported for building IOTA applications. (From Espressif, *ESP Audio DevKits*, Espressif Systems, 2021, www.espressif.com/en/products/devkits/esp-audio-devkits)

16.5.2 Data Structure

The data structure defines how data is systematically stored in the computer for efficient use. In the context of DLT, the data structure is the method of appending approved transactions to the network. Blockchain adds valid blocks using a sequential hash graph (Zhixiong, 2020). Only one block is added at a time, as illustrated in Figure 16.2. This method is not efficient and accounts for blockchain's scalability issues. For instance, statistics shows that the Bitcoin blockchain performs an average of seven transactions per second (Kenney, 2019).

IOTA, on the contrary, implements the DAG data structure. It allows the addition of multiple blocks simultaneously, thereby building the Tangle, as shown in Figure 16.3. This method is highly scalable in comparison with the blockchain. IOTA processes 250 transactions per second on average.

16.6 SUMMARY, CONCLUSION AND RECOMMENDATION

This chapter considers two out of several DLTs in the market: Blockchain and IOTA. It briefly analyzed the architecture and application domains. The focus, however, is

TABLE 16.1
Comparison of Blockchain and IOTA

S/N	Description	IOTA Tangle	Blockchain
1	Consensus protocol	MCMC (coordinators) plus tiny PoW	PoW
2	Data structure	Direct acyclic graph	Sequential
3	Speed (TPS)	250	7
4	Computing resources	Low	Very high
5	Microtransactions	Support	Does not support
6	Transaction charges	Fee free	High transaction charges
7	IoT devices	Suitable	Unsuitable
8	IoT supported hardware	Available	Not available
9	Token volatility	Highly volatile	Moderately
10	Payment medium	Not suitable	Suitable

Ahi, A., & Singh, A.V. (2019). Role of Distributed Ledger Technology (DLT) to Enhance Resiliency in Internet of Things (IoT) Ecosystem. 2019 Amity International Conference on Artificial Intelligence (AICAI), 782–786.

on IoT applications. Technically, attention is drawn to blockchain's and IOTA's architectural features: their consensus protocol and data structure, to be exact. Blockchain technology adopts PoW, PoS, Byzantine fault-tolerant, and others as the consensus protocol. These methods require miners, attracting enormous computation resources as well as high network fees. On the contrary, IOTA implements consensus using MCMC plus tiny PoW. This approach does not require miners, nor does it consume high computation power. Above all, it is fee free.

Furthermore, blockchain implements a hash-graph data structure that sequentially attaches one block at a time. This results in high network overhead, high computing overhead, high storage overhead and low throughput (see detailed summary in Table16.1). The limitations account for the scalability bottleneck in blockchain technology. However, IOTA adopts DAG, an enhanced data structure, to address the scalability challenge of blockchain. In other words, IOTA is relatively highly scalable, fast, high throughput, and fee free. Moreover, it is designed explicitly for IoT with low resource constraints.

In conclusion, owing to the prominent IOTA features outlined, it stands out as a better IoT application option. It is noteworthy that the volatility of the IOTA token renders it inappropriate for a business transaction. This drawback could be a research prospect.

REFERENCES

A, S. (2018). Copyright in the blockchain era: Promises and challenges. *Comput Law Secur Rev*, *34*(3), 550–561.

Ashton, K. (2009). That "Internet of Things" Thing. *RFID Journal*, *22*, 97–114.

Bansal, B., & Rana, S. (2017). Internet of Things: Vision, Applications, and Challenges. *International Journal of Engineering Trends and Technology (IJETT)*, *47*(7), 380–384 May 2017. Seventh sense research group

Bashir, I. (2018). *Mastering Blockchain: Distributed Ledger Technology, Decentralization, and Smart Contracts Explained.* Birmingham, UK: Packt Publishing Ltd.

Behr, A., Thieme, W., Thieme, W., Hepp, M., & Nehme, A. (2020). *Behr Technologies.* Retrieved 02 03, 2021 from https://behrtech.com/solutions/

Cullen, A., Ferraro, P., Sanders, W., Vigneri, L., & Shorten, R. (2021). Access control for distributed ledgers in the Internet of Things: A networking approach. *IEEE Internet of Things Journal.*

Dabeesing, T. (2017, June 28). *Internet of Things (IoT): Regulatory Aspects.* Retrieved 02 03, 2021 from https://www.itu.int/en/ITU-D/Regional-Presence/Africa/Documents/IoT-ICTA-trilok-Dabeesing.pdf

Dorri, A., Kanhere, S.S., & Jurdak, R. (2017). Towards an optimized blockchain for IoT. In 2017 IEEE/ACM Second International Conference on Internet-of-Things Design and Implementation (IoT), Pittsburgh, PA.

Engelhardt, M.A. (2017). Hitching healthcare to the chain: An introduction to blockchain technology in the healthcare sector. *Technology Innovation Management Review, 7*(10), 22–34.

Espressif. (2021). *ESP Audio DevKits.* From Espressif Systems: https://www.espressif.com/en/products/devkits/esp-audio-devkits

Foundation-IOTA. (2021). https://docs.iota.org/docs/iot/0.1/introduction/overview. Germany: IOTA Foundation.

Gde Agung, A.A., & Handayani, R. (In press). Blockchain for the smart grid. *Journal of King Saud University: Computer and Information Sciences.* https://doi.org/10.1016/j.jksuci.2020.01.002

Gupta, S., & Vyas, S. (2021). IoT in green engineering transformation for smart cities. In *Smart IoT for Research and Industry* (pp. 121–131). Cham: Springer.

Gupta, S., Vyas, S., & Sharma, K.P. (2020, March). A survey on security for IoT via machine learning. In 2020 International Conference on Computer Science, Engineering and Applications (ICCSEA) (pp. 1–5). IEEE.

Haber, S., & Stornetta, W.S. (1991). How to time-stamp a digital document. In S.A. Vanstone, & A.J. Menezes (eds.), *Advances in Cryptology-CRYPTO' 90. CRYPTO 1990. Lecture Notes in Computer Science* (Vol. 537, pp. 437–455). Berlin, Heidelberg: Springer. https://doi.org/10.1007/3

Han, K., Liu, S., Zhang, D., & Han, Y. (2012). Initially researches for the development of SSME under the background of IoT. 2012 International Conference on Applied Physics and Industrial Engineering, *Physics Procedia* 24, 1507–1513.

Hyvarinen, H, Risius, M, & Friis, G. (2017). A Blockchain-based approach towards overcoming financial fraud in public sector services. *Business & Information Systems Engineering, 59*(8), 441–456.

IOTA-Foundation. (n.d.). *IOTA Industries.* Retrieved January 29, 2021, from https://www.iota.org/solutions/industries

Jabraeil Jamali, M.A., Bahrami, B., Heidari, A., Allahverdizadeh, P., Norouzi, F. (2020). IoT architecture. In *Towards the Internet of Things. EAI/Springer Innovations in Communication and Computing* (pp. 9–21). Cham: Springer.

Kenney, L. (2019, January 30). *The Blockchain Scalability Problem & the Race for Visa-Like Transaction Speed.* Retrieved February 01, 2021, from https://towardsdatascience.com/the-blockchain-scalability-problem-the-race-for-visa-like-transaction-speed-5cce48f9d44

Kim, H.M., & Laskowski, M. (2018). Toward an ontology-driven blockchain design for supply-chain provenance. *Intelligent Systems in Accounting Finance and Management, 25*(1), 18–27.

Kljajic, D. (2018, May 30). *Blockchain Vs Iota.* Retrieved 02 01, 2021 from https://medium.com/@dakljajic/blockchain-vs-iota-57428da0eb6

Konezny, R., Loch, J., Ueland, M., Riley, K., Marks, G., Ericson, S., Kirkendall, B., Sampsell, D., Roberts, T., & Schneider, T.. (2021). *A Comparison of LPWAN Technologies.* Digi International, Retrieved 02 03, 2021 from https://www.digi.com/blog/post/lpwan-technology-comparison#:~:text=Its%20an%20open%20standard%20operating,radio%20provider%20can%20use%20it

Novo, O. (2018). Blockchain meets IoT: An architecture for scalable access management in IoT. *IEEE Journal of Internet of Things, 14*(8), 1–12.

O'Dair, M, & Beaven, Z. (2017). The networked record industry: How blockchain technology could transform the record industry. *Strategy Change Brief Entrep Finance, 26*(5), 471–480.

Pawczuk, L., Holdwsky, J., Massey, R., & Hansen, B. (2020). *Deloitte's 2020 global blockchain survey.* Retrieved October 23, 2020, from https://www2.deloitte.com/content/dam/insights/us/articles/6608_2020-global-blockchain-survey/DI_CIR%202020%20global%20blockchain%20survey.pdf

Petiwala, F.F., Shukla, V.K., & Vyas, S. (2021). IBM Watson: Redefining artificial intelligence through cognitive computing. In *Proceedings of International Conference on Machine Intelligence and Data Science Applications* (pp. 173–185). Singapore: Springer.

Radanovic, I., & Likic, R. (2018). Opportunities for the use of blockchain technology in medicine. *Applied Health Economics and Health Policy, 16*(5), 583–590.

Salimitari, M., & Chatterjee, M. (2018). A survey on consensus protocols in blockchain for IoT networks. arXiv:1809.05613

Sethi, P., & Sarangi, S.R.. (2017). Internet of Things: Architectures, protocols, and applications. *Journal of Electrical and Computer Engineering, 2017*, 9324035, 25.

Shorten, R., Buchanan, B., & Popov, S. (n.d.). *Overview of IOTA Network.* IOTA Foundation, Retrieved January 27, 2021, from https://docs.iota.org/docs/getting-started/1.1/networks/overview

Sun, J., Yan, J., & Zhang, K.Z.K. (2016). Blockchain-based sharing services: What blockchain technology can contribute to smart cities. *Financial Innovation, 2*(1), 1–9.

Vigneri, L., & Welz, W. (2020). On the fairness of distributed ledger technologies for the Internet of Things. In IEEE International Conference on Blockchain and Cryptocurrency (ICBC), Toronto, ON, Canada.

Xu, M., Chen, X., & Kou, G. (2019). A systematic review of Blockchain. *Financial Innovation*, Springer, *5*(27). https://doi.org/10.1186/s40854-019-0147-z

Yehia, L., Khedr, A., & Darwish, A.. (2015). Hybrid security techniques for Internet of Things healthcare applications. *Advances in Internet of Things, 5*(3), 21.

Zennaro, M. (2016). Intro to internet of things. In ITU ASP COE Training on "Developing the ICT Ecosystem to Harness IoTs, Bangkok, Thailand.

Zhao, H., & Johnson, M. (2021, 01 15). *Global Information Infrastructure, Internet Protocol Aspects, Next-generation Networks, Internet of Things, and Smart Cities.* International Telecommunication Union, Retrieved February 03, 2021 from https://www.itu.int/rec/T-REC-Y/en

Zhixiong, Y. (2020, June 05). *How can the Blockchain/DLT with a DAG data structure ensure immutable transactions without duplicating its transaction across the network or nodes?* Retrieved 02 February 2021, from https://www.researchgate.net/post/How-the-Blockchain-DLT-with-a-DAG-data-structure-can-ensure-immutable-transaction-without-duplicating-its-transaction-across-the-network-or-nodes

Index

C

D

E

F

G

H

R

S

T

U

V

W

X

Y

Z

Printed in the United States
by Baker & Taylor Publisher Services